THE ORIGIN AND DISTRIBUTION
OF BIRDS IN COASTAL ALASKA
AND BRITISH COLUMBIA

THE ORIGIN AND DISTRIBUTION OF BIRDS IN COASTAL ALASKA AND BRITISH COLUMBIA

The Lost Manuscript of Ornithologist Harry S. Swarth

Edited by Christopher W. Swarth

Oregon State University Press   Corvallis

Cataloging-in-Publication Data is available from the Library of Congress.

♾This paper meets the requirements of ANSI/NISO Z39.48-1992 (Permanence of Paper).

First published in 2022 by Oregon State University Press

Printed in the United States of America

Cover artwork: Young (foreground) and adult Golden-crowned Sparrow (*Zonotrichia coronata*) on their nesting grounds near the summit of Monarch Mountain, four miles south of Atlin, British Columbia. In June 1924, Allan Brooks and Swarth discovered four Golden-crowned Sparrow nests there, all with eggs. These were among the first documented and described nests of this little-known sparrow (Swarth 1926). Painting by Allan Brooks. Source: *University of California Publications in Zoology* 30, no. 4 (1926): 51–162.

Oregon State University
OSU Press

Oregon State University Press
121 The Valley Library
Corvallis OR 97331-4501
541-737-3166 • fax 541-737-3170
www.osupress.oregonstate.edu

Contents

Foreword

BY DANIEL D. GIBSON

UNIVERSITY OF ALASKA MUSEUM (RET.)

Harry Swarth was a prolific field and laboratory ornithologist, who over forty-plus years produced a broad and eclectic bibliography of some 220 publications. His works importantly include the descriptions of thirty avian subspecies (seven from California, seven from Baja California, six from the Galapagos Islands, three from Arizona, and three from Oregon), including one from northern British Columbia (Brewer's Sparrow, *Spizella breweri taverneri* [Swarth and Brooks 1925], from Spruce Mountain, 16 km (10 mi) east of Atlin); two from southeastern Alaska (Hairy Woodpecker, *Dryobates villosus sitkensis* [Swarth 1911], from Etolin Island, Alexander Archipelago; and Sooty Grouse, *Dendragapus fuliginosus sitkensis* [Swarth 1911], from Kupreanof Island, Alexander Archipelago); and one from northwestern Alaska (Willow Ptarmigan, *Lagopus lagopus alascensis* [Swarth 1926], from the Kobuk River Delta). His 1920 assessment of the subspecies of the Fox Sparrow (*Passerella iliaca*) remains the definitive work on the subject (Swarth 1920). And his monograph on the birds of Nunivak Island, in the Bering Sea, established a starting point for all subsequent ornithological investigations in that region. His bibliography (Linsdale 1936) includes fifty publications that bear on the ornithology and mammalogy of Alaska and British Columbia.

Swarth left an enduring legacy for natural historians in British Columbia and Alaska and throughout the West. His fieldwork, taxonomic evaluations, and subsequent publications from the late nineteenth century through the first three dozen years of the twentieth continue to provide a foundation that defines much of what is understood (and in some cases what is still not well known) about geographic ranges and relationships among avian species and subspecies throughout northwestern North America.

In finding and resurrecting this eighty-year-old manuscript by his grandfather, and in then transcribing that handwritten material to bring it to light as a published book, Christopher Swarth has provided a serendipitous postscript to Harry Swarth's life and bibliography.

Acknowledgments

I thank Dan Gibson for encouraging me to publish my grandfather's manuscript, and for writing the foreword and preparing the glossary. Thanks to Phil Unitt and Steve Heinl for their thoughtful essays. Steve Heinl read the entire manuscript and his comments improved it and helped me gain a better understanding of bird distribution along the Alaska coast where he lives. Marilyn Fogel and Judith Hurley read early drafts and provided many helpful comments. John Schoen and Rebecca Sentner of Alaska Audubon, Mark A. Chappell, Sam Beebe, and Gavin McKinnon let me use their superb photographs. Anne Fuller provided several family photographs from her collections. Michelle Koo at the Museum of Vertebrate Zoology Archives, University of California, Berkeley, scanned the historical photographs in their collections. Cherie Northon and Thom Eley of Mapping Solutions, Anchorage, redrafted expedition maps and provided excellent cartography. Finally, I appreciate the guidance provided by Kim Hogeland, Marty Brown, and Micki Reaman at Oregon State University Press and the careful copyediting by Laurel Anderton.

Preface

BY CHRISTOPHER W. SWARTH

Southeastern Alaska is a land of rugged, snow-covered mountains, deep fjords, myriad islands in the Alexander Archipelago, and expansive coniferous forests. More than fifty glaciers originate in the Coast Range and terminate in tidal waters. Annual precipitation along the coast is extremely high, creating one of the largest temperate rain forests in the world. To the east, the Coast Range rises to 4,600 m (15,325 ft), marking the boundary between Alaska and British Columbia, and farther inland are vast forests of spruce and pine, numerous lakes, and broad river valleys. At the turn of the twentieth century a more remote and unexplored wilderness hardly existed anywhere else in North America. The Alaska Panhandle and the Coast Range were also the last places on the continent to cast off a mantle of Pleistocene glacial ice and snow that covered this region off and on for 2.5 million years. The region would have been largely devoid of birdlife until climatic conditions moderated and the environment changed in ways that could support higher animals. About ten thousand years

Admiralty Island, southeastern Alaska. Photograph by John Schoen.

ago, as the climate warmed at the close of the ice ages, rocks became exposed and soils developed, allowing plant communities to become established, and eventually, habitats could be occupied by mammals and birds.

The manuscript presented in this book was completed in 1935 by Harry S. Swarth. In it he describes the avifauna of these remote mountainous lands and puts forth ideas about when various species arrived and where they came from once the immense cover of ice had melted and disappeared. This historic manuscript is also a treatise on the distribution and status of the birds of southeastern Alaska and western British Columbia based on the accumulated knowledge up to the 1930s. When Swarth began studying bird communities in coastal Alaska in 1909, only a handful of other ornithologists had conducted fieldwork in this region. Published reports were few and gaps in knowledge were understandably large. The scientific "landscape" was wide open for Swarth to initiate a plan of work on avian diversity, species variation, and the origins of source populations. Knowledge that he gained through a series of expeditions placed Swarth in a unique position to formulate theories to explain the current diversity and distribution of species. In the introduction to his manuscript, Swarth states that the "facts and theories" he writes about are based almost entirely on his own fieldwork, carried out over decades, and on the study of bird specimens he and others collected, which are preserved in California museums where he worked as a research scientist. A digital version of the original manuscript is available at https://archive.org/details/northwesternbird00swar.

Swarth was curator of birds at the Museum of Vertebrate Zoology (MVZ) at the University of California, Berkeley, for most of his career. This position involved both museum and field research. One region of intensive research was the Pacific Northwest, where Swarth led five expeditions: to Alaska's Alexander Archipelago (1909); and, in British Columbia, to Vancouver Island (1910), the Stikine River (1919), the Skeena River (1921), and the Atlin area (1924). Each expedition lasted three to five months, and together they totaled more than 1,100 days in the field. His work was staged from rough field camps in the mountains and on the edge of the ocean. He explored hundreds of miles along the coast in a small, gas-powered boat. A single field assistant accompanied him on each expedition. In 1909, Allen E. Hasselborg, the "Bear Man of Admiralty Island" (see Howe 1996), skippered their boat among the countless islands and deep fjords of the Alexander Archipelago. Hasselborg was an accomplished hunter whose other task was to collect large mammals. In 1910, E. Despard, another bear hunter, traveled with Swarth across Vancouver Island. There, the two ascended a mountain 760 m (2,500 ft) high above Great Central Lake in search of White-tailed Ptarmigan that lived near the summit. Several weeks on Vancouver Island were also spent with Annie Alexander, benefactress and founder of MVZ.

Joseph Dixon, a seasoned Arctic explorer, mammalogist, and photographer, was
with him on the Stikine River (1919). Dixon was also an MVZ research scientist
and had participated in expeditions to the Aleutian Islands, St. Lawrence Island,
Siberia, and Point Barrow. An undergraduate student at UC Berkeley, William
Duncan Strong, accompanied Swarth on the Skeena River expedition (1921).
He was the least experienced field assistant but proved to be an excellent com-
panion and collector. On the 1924 Atlin trip, Major Allan Brooks, a renowned
bird artist, ornithologist, and friend, was his field partner.

Swarth kept an ambitious daily schedule in the field: hiking, observing,
collecting birds and mammals, carefully recording the day's events in a field
journal (a scientific diary), and preparing specimens. Gathering field data
in a remote wilderness required plenty of advance preparation, sticking to a
routine day after day, and perseverance in the face of bad weather— attributes
that can be appreciated by reading his field journals. Except for travel between
campsites, he and his companions worked every day of the week for months
at a time. Shortly after dawn each day, the dozens of small mammal traps that
had been set the night before had to be checked and captured animals carried
back to camp for later dissecting and preparing as study skins. The bulk of the
day was spent covering as much ground as possible and trekking to key locales
where birds could be observed and collected for research. Early twentieth-
century field ornithology almost always involved shooting or "collecting"
specimens because these were needed for careful study in the hand. Collecting
a bird was not easy. Patience, skill, and physical endurance were prerequisites,
first to discover and identify a certain bird, then to get in position and carefully
aim the shotgun, and finally, to recover the bird from the bushes or trees where
it fell. Into the night, in a cabin or a canvas tent, Swarth skinned the specimens
and recorded the day's activities and observations in his journal.

Although Swarth's manuscript was written more than eighty years ago,
his narrative and ideas are still relevant and informative today for anyone who
wants to learn more about the birdlife of the region. Swarth writes, "It's my
hope that this publication might be used by readers to whom those concepts
(life zones and faunal areas) are unknown or only vaguely understood." With
that goal in mind, he wrote his manuscript to be engaging and accessible, not
like a technical report meant for other specialists. Swarth plowed new scientific ground by thinking and writing about distributional ranges, colonization,
source populations, and morphological variation in those populations that
arose over millennia when separated by geographic barriers. He wanted to
understand the complex patterns of diversity and distribution, species and
subspecies formation, and the shifts in geographic ranges of birds and mammals that were taking place following the end of the Pleistocene epoch. He

Prince of Wales Island complex, southeastern Alaska. Photograph by John Schoen.

recognized that the thousand-mile-long Coast Range was a major impediment to the movements and migrations of animals between the interior of British Columbia and the Pacific coast. It was not common then, as it is now, for an ornithologist to consider how zoogeography and evolution interact to create regional species diversity. Swarth was among the first to apply the concepts of life zones and faunal areas to the region, theories recently developed by zoologists C. Hart Merriam and Joseph Grinnell. His thorough documentation of the entire avifauna also established a starting point for ornithologists who monitor population changes today in this part of the Pacific Northwest. The key publications that provided the foundation for the manuscript are shown in bold print in Swarth's bibliography on page 149.

SPECIES, SUBSPECIES, AND COLONIZATION

Much of Swarth's manuscript is devoted to comparing the bird species in the coastal or Sitkan district with related birds found in the Cassiar and Omineca districts east of the Coast Range. Most of us are aware that a species is a fundamental unit of biological classification, but we may be less familiar with the concept of a subspecies. Many species that are broadly distributed, whether animal or plant, are made up of geographic races or subspecies. The American Ornithologists' Union (AOU) *Check-list of North American Birds* (1998) states that "subspecies names denote geographic segments of species' populations that differ abruptly and discretely in morphology or coloration." In biology, a subspecies is given a three-part name, or trinomial, that consists of genus, species, and subspecies. The Northern Flicker, for example, includes the "Yellow-shafted" subspecies (*Colaptes auratus luteus*) found primarily in eastern North America, and the "Red-shafted" subspecies (*C. a. cafer*) in the West. Where these subspecies meet, as Swarth discovered near Hazelton, British Columbia, individual flickers with characteristics of both subspecies can occur. Similarly, throughout the Pacific Northwest, there are two Dark-eyed Junco subspecies. The "Oregon" junco has a blackish hood, occurs along the immediate coast, and is the common junco of the Sitkan district. The "Slate-colored Junco" has a grayish hood and darker back and occurs inland of the Coast Range. These junco subspecies comprise distinct regional populations that exhibit observable geographic variation in appearance. They rarely occur together during the breeding season and can be easily identi-fied in the field any time of year. The identification and description of isolated populations and subspecies can be very important for conservation and popula-tion monitoring. Some subspecies are considered to be "species in the making." According to Kevin Winker (2010), "The study of subspecific variation in birds has been an important driving force in the development of evolutionary biol-ogy." He goes on to say that "perhaps the most useful academic purpose to which subspecific variation has been put is as a frame of reference for posing questions about animals and evolution."

Many of Swarth's ideas have stood the test of time. Byron Weckworth et al. (2005) write, "Since Swarth's (1936) characterization of the Sitkan District, the North Pacific coast has been recognized as a distinctive biogeographical region in North America." According to Joseph A. Cook et al. (2006), "Swarth . . . provided a baseline for understanding biotic diversity in the region, which is largely isolated from the remainder of North America by the coastal moun-tains." Recent archaeological and molecular studies have determined that some western parts of the Alexander Archipelago may not have been glacier covered, and that these glacier-free areas, called "refugia," could have allowed some plants and mammals, and perhaps even some birds, to survive there (MacDon-ald and Cook 1996; Cook et al. 2006). In explaining the unusual distribution of Spruce Grouse (*Falcipennis canadensis isleibi*) in southeastern Alaska, Dicker-man and Gustafson (1996) speculated that this bird—found only in the Prince of Wales Island complex—could be a relict of glacial refugia that existed during the Pleistocene. Their theory is based on forty-thousand-year-old grizzly bear and marmot bones discovered on Prince of Wales Island. Genetic studies of the "Prince of Wales" Spruce Grouse support this theory (Barry and Tallmon 2010). Colonization of this region by birds was probably more complicated than Swarth envisioned in the 1930s (Heinl and Piston 2009).

ECOSYSTEM CHANGES OVER A CENTURY

How has the ecosystem of southeastern Alaska and northwestern British Columbia changed over a century? This region is still an immense, sparsely populated wilderness, as it was when Harry Swarth explored it. The towns he visited in northern British Columbia are hardly larger now than they were then. While many things have changed in this region of the Pacific Northwest, much remains the same. The short-lived Klondike Gold Rush between 1896 and 1899 had come and gone by his time. Small towns swelled, then abruptly shrank back to pre-gold-rush size. The town of Atlin, for example, grew to ten thousand inhabitants in 1898 but today has a resident population of only five hundred. It could take Swarth a week or longer to travel from Berkeley to Atlin in the 1920s and 1930s, depending on ice cover on the lakes near Carcross, Yukon. From Berkeley he rode the train to Seattle or Vancouver, where he boarded an ocean steamer bound for Skagway. From Skagway the White Pass and Yukon Route Railway carried him the sixty miles over the mountains to Carcross. From Carcross a steamship carried passengers down Tagish Lake to the portage at Taku for the ride on the Taku Tram railroad (at only 4 km, the shortest one in Canada) to the edge of Lake Atlin, where another steamer completed the trip to the town. Today, Atlin is easily reached by two highways and is a popular summer vacation destination. While northern British Columbia is still thinly populated, seventy-three thousand people now live in southeastern Alaska, about 10 percent of the state's population, and cruise ships bring a million tourists every summer. Although shipborne tourism is a boon to the economy, the trade-off can be degradation of coastal air and water quality.

The impacts of logging, mining, and climate change are also being felt. Of primary importance to the region is the Tongass National Forest, the largest national forest in the United States. It encompasses about 80 percent of all the land in southeastern Alaska and contains a major proportion of the world's remaining old-growth temperate rain forest. The ecological health of the Tongass is crucial to the regional economy and environment. According to the *Ecological Atlas of Southeast Alaska* (Smith 2016), "The rich rainforest landscape is the primary reason why communities and industries have thrived on the Tongass forest for so long. With the exception of mining, the resource-based industries of commercial fishing, cruise ship tourism, and timber depend on intact, healthy forest. This is also true for subsistence hunting, sport fishing, bird watching, and many other human use aspects of Southeast Alaska." Logging operations were relatively small in Swarth's day and up until the 1950s were still carried out by hand, with individual trees selected for cutting. Today industrial-scale logging uses heavy equipment to clear-cut the forest, leaving huge swaths devoid of vegetation in its wake. The forests of Vancouver Island have been extensively logged by clear-cutting. The most serious environmental issues in this entire region are often attributed to large-scale logging, which has removed, for example, about a third of the trees in the Tongass. Not only does logging directly impact the surrounding environment where it occurs, but the associated wider land disturbance and road building have ripple effects on the forest and on the regional economy, including commercial and recreational fishing and tourism. For many decades the cutting of old-growth trees in the Tongass has been controversial and the subject of considerable litigation. An old-growth forest is one that exhibits a stable, climax state, consists of trees that are over 150 years old, and has not been directly impacted by humans. It is estimated that the old-growth forests in the Tongass and on Vancouver Island can store over one thousand tons of carbon per hectare, which is one of the highest rates on earth. Their loss thus contributes significantly to global climate change. Old-growth forests also support diverse bird communities. In 2020, the US Forest Service and the Trump administration tried to roll back the federal Roadless Area Conservation Rule of 2001 to allow logging, new roads, and development on nine million acres of forest (much of it consisting of old-growth trees) that had long been off-limits in the Tongass. Road building can be as damaging to the forest as clear-cutting and is often a precursor to further development. Fortunately, indigenous people and conservation organizations convinced the current federal administration to scuttle the Trump-era plans to quash the Roadless Rule. Industrial-scale logging has declined significantly since the 1990s, with smaller companies practicing selective logging once again.

In southeastern Alaska, large silver and gold mines are located near Juneau and on Admiralty Island, but mining is more extensive in northwestern British Columbia. Serious environmental impacts on aquatic environments have long been associated with the runoff and tailings from mining operations. In the Pacific Northwest, damage from mining to salmon spawning areas is a great concern to many. British Columbia mines can also impact Alaska waters if contaminants such as heavy metals and acid drainage enter transboundary rivers that flow west through the Coast Range and into Alaska. Long-abandoned gold and copper mines in British Columbia continue to impair water quality decades after operations ended.

Climate change is another significant concern in the region, having a disproportionate impact on environments in northern latitudes. A dramatic and visible result of climate change in southeastern Alaska is the retreat of tidewater glaciers. Today, 95 percent of Alaska's one hundred thousand glaciers are thinning or retreating. According to the National Park Service, there is 11 percent less glacial ice in Glacier Bay today than in the 1950s. When Captain Cook arrived at

Upper Tenakee Estuary, Baranof Island, southeastern Alaska. Photograph by John Schoen.

the mouth of Glacier Bay in 1778, the Grand Pacific Glacier was 1,220 m (4,000 ft) thick and projected into Icy Strait. Today, this glacier has retreated far inland, opening up a 105 km (65 mi) waterway that did not exist two hundred years ago. Over the past century in southeastern Alaska, mean winter temperatures have risen by 2°C and summer temperatures by 1°C. Climate change projections indicate this trend will continue, with shorter, warmer, wetter winters and longer, hotter, drier summers. Predicted outcomes from increasing air temperature include faster glacial melt, an elevational shift in the snow line and related reduction in snowpack, a change in precipitation from snow to rain below that line, increases in stream temperature, and effects on the distribution and abundance of wildlife because of underlying changes in habitat.

CHANGING BIRD POPULATIONS

Given these many changes and impacts, how have bird populations changed over a century? Based on a recent comprehensive analysis, bird species across North America have declined dramatically since 1970, with losses estimated at a staggering 2.9 billion birds, representing 303 of 529 species (Rosenberg et al. 2019). These depressing statistics are all related to human activity: habitat loss and degradation, agricultural pesticides, deforestation, direct anthropogenic mortality, and invasive and alien species such as cats and rats—all problems that are exacerbated by climate change. In the region Swarth studied, habitats for birds have become increasingly fragmented and altered by road building and log-ging—especially clear-cutting—since the 1950s. Clear-cuts fill with dense shrub growth, which does provide suitable and temporary habitat for some songbirds. Second-growth forests follow in the process of succession. Old-growth forests, however, support a more diverse fauna than do second-growth coniferous for-ests. Although no bird populations have become extirpated since Swarth's time, a number of species are now vulnerable, including, for example, the Queen Char-lotte Goshawk (*Accipiter gentilis laingi*) and the Prince of Wales subspecies of the Spruce Grouse (*Falcipennis canadensis isleibi*) (Smith 2016). Today natural habi-tats for birds are protected within many national and provincial parks and other natural areas, including, for example, the immense Glacier Bay National Park and neighboring Tatshenshini-Alsek Provincial Park. To raise more awareness about birds and their habitats, BirdLife International and the National Audubon Society established the Important Bird Area (IBA) program in the 1980s, which identifies places of national and international significance for bird populations and conservation. Southeastern Alaska has sixteen IBAs, seven of which exceed 100,100 ha (250,000 ac), and coastal British Columbia has fifty-five IBAs.

Bird ranges expand and contract gradually over time. In the Pacific Northwest, the ranges of breeding birds and the occurrence of rarities are well documented today by legions of birders. The *Atlas of the Breeding Birds of Brit-ish Columbia*, a seven-year project by 1,300 citizen-scientists, has mapped the breeding range of 320 species and is a tremendous resource for understanding provincial avifauna (Davidson et al. 2015). Since 1935 the breeding ranges of some species have apparently expanded westward from the interior of British Columbia and now occur on the coast, especially along some of the major main-land rivers (Johnson et al. 2008; see essay by Steve Heinl). Over forty-five spe-cies that Swarth considered absent or very rare in southeastern Alaska in 1935 have now been sighted, and some are breeding in small numbers or in restricted locales on the coast (Steve Heinl, pers. comm.; Heinl and Piston 2009). In the last fifty years ten new species have become firmly established in southeastern Alaska (Kessel and Gibson 1994; Heinl and Piston 2009). Notable among the new arrivals, all absent in Swarth's day, are the Wood Duck (*Aix sponsa*), Anna's Hummingbird (*Calypte anna*), Caspian Tern (*Hydroprogne caspia*), Barred Owl (*Strix varia*), Eurasian Collared-Dove (*Streptopelia decaocto*), European Starling (*Sturnus vulgaris*), and House Sparrow (*Passer domesticus*). In north-western British Columbia, examples of species confirmed as breeding that were very likely absent in Swarth's day are the Wood Duck, Blue-winged Teal (*Anas discors*), Pacific Loon (*Gavia pacifica*), Least Sandpiper (*Calidris minutilla*), Barred Owl, and Red-winged Blackbird (*Agelaius phoeniceus*) (Campbell et al. 1990–2001; Davidson et al. 2015). In 2020, Audubon Alaska and collaborators completed the Southeast Alaska Birding Trail, a virtual guide (not an actual trail) to eighteen key places for bird-watching, which will certainly lead to even further knowledge of bird status and distribution.

In 1935, Swarth passed away before his manuscript could be published. Harry Swarth was my grandfather, and in 2016 I discovered the manuscript in a box of family documents. As I read through it and began transcribing the handwritten pages, I realized that the manuscript would be a very useful contribution to present-day knowledge of Pacific Northwest birdlife and should be published.

PLAN OF THE BOOK

The first section consists of Swarth's original manuscript, transcribed and presented exactly as written in 1935. His introductory chapters describe the geomorphology, forests and plant communities, and avifauna, from the Sitkan district of southeastern Alaska across the Coast Range to the Cassiar and Omineca districts of northwestern British Columbia. A synopsis or "Accounts of the Various Species under Consideration" follows, focusing on the distribution, abundance, and systematic relations of over 195 species and subspecies.

The second section includes expedition maps and itineraries, field camp locations, field journal entries, and many historic photographs. Journal entries describe travel, logistics, and the occurrence and behavior of the birds he encountered. These entries also reveal the difficulties of the day-to-day work of a field ornithologist in the early twentieth century as well as the excitement of new discoveries. Photographs, most of them taken by Swarth, reveal the stark and foreboding yet beautiful landscape.

The third section includes bird species checklists that I compiled from Swarth's field journals, which show the diversity of birdlife in many of the areas where he worked.

The fourth section contains essays that take up aspects of Swarth's research to show how his work still influences current ornithological studies. Steve Heinl describes similarities and differences in coastal birdlife between Swarth's day and today. Migration and variation in western Fox Sparrows are discussed by Phil Unitt. In a third essay, I recount the tortured taxonomic history of the Timberline Sparrow, discovered by Swarth and Allan Brooks on the 1924 Atlin expedition, and the unresolved controversy about its status as a full species.

The fifth section is a glossary that will allow readers to match the historic, out-of-date bird names used in the book with the modern common and scientific equivalents.

The sixth section is a bibliography of Swarth's publications on the birds and mammals of the Pacific Northwest.

Biography of Harry S. Swarth

Harry Swarth was born in Chicago in 1878 and developed an early interest in birds. With his family he would visit the Chicago Field Museum, where he first became aware of museum specimen collections. In 1886, the family moved to Los Angeles, where his passion for birds grew as he roamed freely, exploring the open spaces and diverse birdlife of southern California. His interests were encouraged by his parents and by family friend and mentor George Frean Morcom. Harry learned to prepare bird study skins by watching Frank Stephens, curator of the San Diego Natural History Museum. The Cooper Ornithological Club had formed in 1893, and Swarth, fifteen at that time, soon became a member. The club was essential for bringing bird-watchers and ornithologists together in California, and through his association and later leadership in the club, Swarth made lifelong friends and met professional colleagues.

In 1896, when Swarth turned eighteen, he organized and led a remarkable ornithological expedition with three other young men to Arizona to

Harry Swarth preparing specimens in the southern Sierra Nevada, Fresno County, California, August 1916. Photograph by Joseph Dixon, with permission of the Museum of Vertebrate Zoology, University of California, Berkeley.

1

explore and study the birdlife of the Huachuca Mountains. In February of that year, the teenagers headed southeast with two horses and a wagon filled with camping and collecting gear—walking 1,050 km (650 mi) from Los Angeles to southern Arizona. Along the way and while camped in Ramsey Canyon (a current mecca for birders), they observed 169 bird species and collected over 400 specimens for museums back in California. Harry Swarth's field notes from this first research expedition were recently discovered and published (C. Swarth 2018). His investigations in Arizona continued for the next twenty years, alternating with field work in the Pacific Northwest. Fifteen publications resulted from his Arizona work, the first being *A Distributional List of the Birds of Arizona*, published in 1914 by the Cooper Ornithological Club.

In 1908, zoologist Joseph Grinnell had just become the first director of the new Museum of Vertebrate Zoology (MVZ) at the University of California, Berkeley. MVZ was fast becoming a center for the study of vertebrate systematics and evolution in the West, and today it is one of the premier zoological research museums in the world. Swarth, who by that time in 1908 had returned to Chicago to work in the Zoology Department at the Field Museum, was contacted by Grinnell and hired to become MVZ's first research zoologist. In 1910, he was promoted to curator of birds, a position he held nearly continuously until 1927. He lived in Berkeley with his wife, Winifern, and they raised two sons, George and Morton.

At MVZ, Swarth combined fieldwork, collecting, and specimen-based museum studies to investigate myriad questions about avian speciation, migration, systematic relations in birds and mammals, geographic distribution, and morphological and plumage variation within species. His research was concentrated in California, Arizona, the Pacific Northwest, and the Galapagos Islands. A long and detailed monograph on Fox Sparrow subspecific variation, geographic distribution, and migration routes is an especially well-known paper (Swarth 1920). In that paper he introduced the concept of "leapfrog" migration: the northernmost Fox Sparrow subspecies, which nests in Alaska, while migrating south in the fall, "leapfrogs" over more southerly nesting populations (some of which are year-round residents) and travels farther to wintering grounds in southern California. Much of his research over the years dealt with species that were little known or perplexing taxonomically, including geese, grouse, owls, woodpeckers, jays, wrens, waxwings, and sparrows. In 1916, Swarth was elected a fellow of the American Ornithologists' Union, an honor reserved for the fifty foremost ornithologists of the Western Hemisphere.

In 1927, Swarth became curator of the Department of Mammalogy and Ornithology at the California Academy of Sciences in San Francisco, a position he held until his death in 1935. While there, he analyzed the academy's

Harry Swarth, 1934. Swarth family collection.

large collection of Galapagos Island birds (8,700 specimens), which had been collected by Rollo Beck and others on the academy expedition of 1905–1906. Swarth also examined Stanford University's specimens from the Hopkins-Stanford Galapagos expeditions of 1889 and 1899. To acquire a fuller understanding of Galapagos avifauna, Swarth traveled to the British Museum (now the Museum of Natural History at Tring) to examine the type specimens of birds that Charles Darwin had collected on the Galapagos Islands during his around-the-world cruise on the HMS *Beagle*. In 1932 Swarth journeyed to the Galapagos as chief scientist aboard the yacht *Zaca* on the California Academy of Sciences–sponsored Templeton Crocker Expedition. Papers he published from this research (see Swarth 1931, 1934) were the first to place Darwin's finches and other species, pivotal to our understanding of evolution, into a modern systematic and evolutionary framework. Later, Swarth and Robert T. Moore (of the Moore Laboratory of Zoology, Occidental College, California) worked to persuade the government of Ecuador to conserve and protect the Galapagos Islands.

Over the course of his career Swarth published 220 papers and described thirty new birds (nine with Joseph Grinnell) and eleven new mammals. In 2011, the Western Field Ornithologists established the Harry S. Swarth Award in Western Field Ornithology to honor his legacy. Comments written by colleagues after he passed away described him this way:

> It is hard to say which was most impressive—his remarkably accurate recognition of the plumages of certain birds we encountered, or his patience in withstanding, with meager camping equipment, an exceptionally stormy May in the north coast ranges.
>
> . . . Of all his undertakings, Swarth's revision of the fox sparrows (*Passerella iliaca*) was no doubt the most difficult. At the same time, it stands out not only among his own reports, but in general systematic ornithology as a model of organization, clear presentation, and significant interpretation. This publication is probably the best example among his writings of his ability to simplify a complicated problem and then to present it in a logically consistent and concise manner. (Linsdale 1936)

> Swarth had early encountered some of the complications that arose from the study of what were first called "geographic variations" and are now recognized as subspecies. Becoming greatly interested in the variations and the inter-relationship of certain groups, he made an intensive study of them, and his comprehensive work resulted in many papers that have been of great value in the classification of certain forms not only of birds but of mammals as well. (Mailliard 1937)

SECTION 1

The Origin and Distribution of Birds in Coastal Alaska and British Columbia

By Harry S. Swarth (1935)

INTRODUCTION

My theme is found in the avifauna of a section of northwestern North America; it is concerned with the inter-relations and distribution of the birds of the southeastern portion (the "pan-handle") of Alaska and the neighboring interiors of British Columbia. Here is to be seen the curious accident of a political boundary coinciding exactly with a faunal boundary that is as trenchant as any to be found in North America. In this part of the continent Atlantic coast birds press nearly to the Pacific, to points that are almost within sight of the western sea. The Pacific avifauna is correspondingly restricted to an extremely limited coastal area. These facts in a general way are, of course, common knowledge, but there has never been a thoroughgoing presentation of conditions with details of each species and with resulting deductions.

The boundaries that define alike the habitats of large numbers of species, thus forming "faunal areas," the nature of the species concerned, their origin and relationships, the occasional aggressive species that we can see pressing onward past the barrier that halts most of their associates, and the conditions that such pioneers are encountering, all these form the subject-matter of the following pages, presented, I hope, so as to have retained some portion of the interest that obtained in the original observations.

These facts and theories are based mostly upon the results of a series of summer trips that I made, first for the Museum of Vertebrate Zoology of the University of California, then for the California Academy of Sciences. The primary object of these expeditions was the gathering of a series of specimens of birds and mammals for the museum collections, supplemented by such written observations upon the species as lay within the capability of the collector. Long summers were spent upon the coast of southeastern Alaska (1909),

Harry Swarth, ca. 1910. Swarth family collection.

on Vancouver Island (1910), in the Stikine River drainage of British Columbia and Alaska (1919), in the upper Skeena Valley (1921), and in the Atlin region (1924, 1929, 1931, and 1934). The resulting collections, totaling about 4,000 birds and 2,500 mammals, formed the basis for a series of published reports that were concerned mostly with the systematic questions involved.

The distribution of species and subspecies formed an important aspect of those studies, and a feature that was dwelt upon in some detail, but no one of the several publications contains a general summary of the facts of animal distribution in the region, a statement of correlations such as may be lost to sight in the detailed information presented. These generalizations, here presented, should be of interest to a wider circle than that formed by specialists in systematic studies. The subject has not been treated in any general work in ornithology. The observer of birds for the pleasure of it, if he chances to visit this region, and it does attract thousands of tourists each year, can direct his observations to better purpose when he understands the meaning of the many puzzling and seemingly contradictory occurrences that he will encounter. The student of other phases of natural history may be able to apply some of these observations in his own field.

Introductory chapters are descriptive of the Sitkan district of the coast, of the Cassiar and Omineca districts inland, these terms being applied to defined faunal areas, and the descriptive matter covering such aspects of the surroundings obviously affect bird life. The chapters dealing with groups of bird species are each preceded by a list of the species in that group. These lists are in parallel columns, pertaining respectively to coast and interior, with the species or subspecies that are complementary in the two regions placed opposite. The richness or poverty of the region so far as that bird group is concerned, and the condition of possible "ecologic niches," so called, perhaps crowded, perhaps unoccupied, are thus made apparent.

The elementary nature of explanatory statements in the sections dealing with life zones and faunal areas is due to the hope that this publication might be used by readers to whom those concepts are unknown or only vaguely understood.

THE SITKAN DISTRICT: TOPOGRAPHY, CLIMATE, AND PLANT COMMUNITIES OF THE COAST

The "Sitkan district" of southeastern Alaska received that name from E. W. Nelson, who, in 1887, published his "Report Upon Natural History Collections Made in Alaska," in which the physical characteristics and the biological peculiarities of that immense territory were described with such accuracy, at least as to their salient features, as to leave for later observers but little more than the filling in of details. The Sitkan district extends north from Dixon Entrance (the southern boundary of Alaska) at least to Skagway. Nelson's definition of the region carried it on as far as the base of the Alaska Peninsula, but later writers have been inclined to restrict it as indicated. It includes the extremely narrow strip of mainland coast that lies west of the crest of the coastal mountains (of an average width of perhaps thirty or forty miles), together with the numerous islands, large and small (the Alexander Archipelago) that lie to the westward.

The outstanding feature of the region (one that the most casual observer cannot overlook) is the tremendous rainfall. This country comprises the wettest part of the northwestern rain-belt, with the average annual precipitation at different points (there is considerable local variation) ranging from 75 inches to 200 inches, or perhaps even more. Constant rains of summer are followed by heavy snowfall in winter. The blame for this climatic feature may be divided between the Japan current and the mountain barrier that lines the coast. The adjoining sea is relatively warm and the warmth affects the air above. The warm, moisture-laden sea breezes, pressing eastward, are abruptly cooled by the icy summits of the Coast Range and drop their watery content in the all but unceasing misty rain that is so characteristic of the country.

Immediately beyond the mountains (sometimes within 60 or 70 miles of the coast) the rainfall decreases abruptly, in many sections to not more than 10 or 15 inches annually. Overcast skies are not uncommon just east of the Coast Range, and ragged streamers of clouds may extend from jagged peaks, where, laterally, they seem to be caught, and to hang helplessly until dissolved; but what would be threatening weather in most parts of the world may be disregarded here as but a threatening gesture.

The heavy winter snows of the coastal region, presumably in past ages far in excess of what was carried away in the brief seasons of melting, have resulted in the vast deposits of glacial ice with which the mountains are burdened, for southeastern Alaska is one of the several great glacier-covered areas of the earth, second only to the illimitable ice fields of Greenland and Antarctica. It is a well-known fact that still needs repetition for realization: Alaska glaciers are all in the southernmost part of that territory, and mostly well south of the Arctic Circle, while truly Arctic northern Alaska is entirely free of glaciers and has been so for ages. We are told, and the evidence is there for all to read, that the glaciers of southeastern Alaska are but the surviving remnant retreating, year by year, of a vastly greater ice cap that once spread over much of western North America from the center, a glacial covering that was an important factor in scouring out the intricate network of channels that divides the islands of the Alexander Archipelago.

The Coast Range that borders the shores of the Pacific from central California northward culminates in Alaska in a colossal series of jagged peaks and ridges towering abruptly from depths below the water. This part of the range literally is not snow-capped in all its length, but lines of perpendicular cliffs and sharp-pointed, bare, granite summits are thrust upward far beyond the encircling weight of glacial ice and snow that hangs everywhere along the lesser heights, and that clogs valleys and canyons downward sometimes to saltwater. This is mostly on the mainland. The islands, just as mountainous, do not attain such lofty elevations, and whereas the mainland summits from their raw irregularities appear never to have been subjected to the smoothing passage of the grinding ice masses, the islands exhibit mostly a somewhat hummocky effect and gently rounded outlines, forest-covered nearly everywhere. True, on almost every island there are summits that are snow covered perpetually, and on some of the largest such as Baranof and Chichagof, these snowfields stretch for tremendous distances, but, nevertheless, to the traveler on a passing steamer, going up or down the coast, the general effect is of a naked, harsh and impassable barrier to the eastward and of a milder setting of rounded hills and more gentle slopes, smothered with vegetation, on the surrounding islands off the coast.

North of Dixon Entrance the fields of glacial ice that must cover unbrokenly large sections of the unexplored mountain masses send tongues downward to the sea, in increasing volume and frequency as one travels northward. Taku Glacier at the west (near Juneau) and Lewellen Glacier at the east (on Atlin Lake) are but names applied to different corners of one huge field of ice. Great Glacier and Mud Glacier, on the Stikine, and others, unnamed, that may be seen along the intervening coast, are undoubtedly all just projecting extremities radiating from another great center.

The summers in the coastal region are never very warm, nor the winters very cold, a climatic feature that is different indeed from what prevails inland. One striking result of these peculiarities of temperature and rainfall is seen in the characteristic vegetation of the coastal belt, where forest trees and underbrush both attain an exuberance of growth that can be compared only to the jungles of the tropics. Only one who has struggled through these woods can realize how nearly impassable they are. Trees grow on every inch of ground that will support a tree, from mountain summits down to the water's edge. The forest's lower border is sharply defined, marked by one inflexible barrier, saltwater. It is a fact that meadows may be seen here and there back from the beach, sometimes of great areas, grown with tall grass and entirely free of trees, but more than one inexperienced camper, using this easily available campsite, has been awakened at night by the invading saltwater in his tent. It may be only the occasional high spring tide that reaches that level, but one may feel assured that any such open meadow lies under saltwater at some time, however unlikely it may seem.

The forested zone covers by far the greater part of the Sitkan district. The characteristic and dominant tree here, and for many miles to the southward, is the Sitka spruce (*Picea sitchensis*). From tide line upward to timberline over hill and valley, on ridges and along the rivers, with lesser admixtures of hemlock and cedar, this tree grows in such density as to suggest nothing so much as a particularly thick pelaged fur. The view that meets one from a coastal steamer, day after day, is of an interminable series of vistas of dark green slopes covered with almost unbroken stands of conifers, the generally green coloration so thickly speckled with gray dead trunks as to produce almost a pepper-and-salt effect. Under favorable conditions some trees attain a height of 100 or even 150 feet, but ordinarily, in the crowded forest that is most prevalent, the spruce trees are under 100 feet high. Judging from the abundant dead trees they would seem to be not very long lived; perhaps the stony mountainsides or boggy valleys that form most of their habitat are unequal to the support of really fine forest trees.

The western hemlock (*Tsuga heterophylla*), less abundant than the ubiquitous spruce, grows over much of the coastal region. It is most conspicuous away from tide water, thriving in inland valleys and on mountains, sometimes to the exclusion of the spruce trees over considerable areas. Both the western red cedar (*Thuja plicata*) and the yellow cedar (*Chamaecyparis nootkatensis*) are fairly conspicuous features of the Alaska woods, beautiful trees, often of a great size.

Along the beaches, just above high tide and bordering the forest, there is almost everywhere a line of alders (*Alnus oregona* and *A. sitchensis* both occur in this region) growing as dense, rounded bushes, the lower branches close to the ground. Marshy areas may have extensive growths of this bush, mixed with willows, the stiff crinkly branches of the alder twisted and crushed down by winter snows, forming an exasperatingly difficult obstacle to be struggled through. Associated with the alder are thickets of salmonberry (*Rubus spectabilis*), in late summer laden with berries that are acceptable alike to man, bear, and bird, and also the highly objectionable devil's club (*Fatsia horrida*). This broad-leaved bush has thick, fleshy stalks with erect clusters of red berries, the foliage green, glossy, and attractive—at a distance. Close contact reveals a truly diabolical array of thorns, on branches, twigs, and even on the undersides of the leaves, needle-sharp, prone to break off under the skin, and as poisonously painful as cactus spines. Any attempt to traverse the Sitkan forests must take into account the inevitable presence of this frightful shrub, a final straw added to the difficulties of densely crowded trees, sprawling and tangled deadfalls, moss-hidden rock piles, and the persistent hindrance of the closely woven alders—all dripping water in gentle but penetrating streams. An attractive bush, most abundant on the mainland coast, is the red elderberry (*Sambucus racemosa*), its gleaming fruit adding a touch of brilliant color to the generally

Harry Swarth in the field at Doch-da-on Creek, Stikine River, British Columbia, 23 July 1919. Photograph by Joseph Dixon, with permission of the Museum of Vertebrate Zoology, University of California, Berkeley.

rather somber woods. The larger rivers of the mainland, notably the Stikine, are bordered by groves of tall cottonwoods (*Populus trichocarpa*), which grow rapidly on treacherous bottomlands where shifting streams uproot hundreds of them every year and sweep the fallen giants downstream to the coast. There, on island and mainland beaches and tidal flats, the cottonwoods feature as denuded and water-worn logs, thrown up by the sea to form tangled obstructions, in some places for miles along the water's edge.

In all four zones there are, of course, multitudes of other plants, besides the conspicuous ones just mentioned. Broad-leaved skunk cabbage borders the meadows, one of the first bits of green to appear in the spring, and often attaining enormous size, with leaves three or four feet long. A carpet (a very uneven one) of moss, forming soft cushions a foot or more deep, covers much of the forest floor, and masses of ferns flourish in the wet, shaded depths of the woods. Flowers, and some relatively brilliant ones (such as the red "Indian paint brush" and the blue lupine), are found in the more open places, but despite such occasional bits of color, the general effect is somber, the prevailing tone being given by the dark greens of grass and bushes, the dark vistas of the crowded forest, and the dark depths of the still waters round about. The woods are generally hushed and still, seldom disturbed by song of bird or chirp of insect, while the constant drip of falling water may hardly be said to break the general quiet.

In this region there are four types of vegetational surroundings that may be recognized as distinct and characteristic, namely, littoral, forest, tundra, and mountaintop. Above the dripping, rock-clinging kelp, alternatively exposed and covered by saltwater twice a day, are the various saltwater grasses or grass-like plants that endure nearly as much submergence. Then, higher in the littoral zone there are other "grasses" (I use the term very loosely), sometimes waist high or higher. Clumps of willows, alders, elder and high-bush cranberry, tall broad-leaved wild parsnip, tangles of wild peas, and thickets of salmonberry. This zone is relatively open and easy to traverse. In fact, one is usually forced into it by the forest wall crowding alongside.

The intricate channels between the closely studded islands of the Alexander Archipelago wind their tortuous courses for many miles back from the open sea with width and depth varying between wide limits. There are impressive waterways that measure miles from shore to shore, others of lesser size down to narrow runs a hundred yards or less in width, and some that may even be crossed dry shod at low tide. Geologists tell us that these are all "drowned valleys," scoured out by glaciers either before or after submergences below sea level, and indeed at many points the U-shaped cross-section suggestive of such action is plainly to be seen, especially from a steamer traveling in mid-channel.

In scores of places the channels extend inland as fjords, extremely narrow and with steep granite walls with trees clinging most incredibly to dubious root-holds down to high-tide level, below that, down to the water's edge, a breadth of precipitous, polished rock on which a fly could hardly find a foothold. There may be miles of such coastline where a moderate-sized vessel could tie up as to a dock, and, with length of rope to allow for changing tides, be safe as to depth of water though within arm's length of shore. In such a fjord ("canal" is the local term) it is generally useless to seek a landing place until the very head is reached, perhaps 20 or 30 miles inland from the general coastline, and there, where the valley emerges from the sea, is first an expanse of grass-grown flats, intersected by a network of narrow ditches, and, some distance farther back, beyond the reach of the highest tides, the forest, stretching inland to unexplored distances. Any such large fjord will have a good-sized stream, or maybe several, debouching at its head, usually from glacial sources.

Consider that twice in twenty-four hours the changing tides must come up these constricted waterways, not only the ten, twenty, or thirty miles of that channel's course, but the one hundred or two hundred miles farther from where the open sea lies beyond the island barrier, and an understanding is reached of the reason for the tremendous rise and fall of the tides everywhere along the innermost coast. Twenty or thirty feet of variation between high

Boat at Kupreanof Island field camp, southeastern Alaska, 1909. Photograph by Harry S. Swarth, with permission of the Museum of Vertebrate Zoology, University of California, Berkeley.

and low tides is not unusual here, where a vast volume of water charges inland between increasingly constricting walls. Steamers at any of the ports are tied to the docks by cables that have to be continually re-adjusted, and it is an ordinary experience for a shore-going passenger, stepping off perhaps on a gang-plank that slopes steeply upward from a steamer deck to wharf, to find conditions reversed on his return an hour or so later and the same gang-plank sloping as steeply upward from the wharf to the boat's deck.

The sheltered canals and sounds of the Alaska and British Columbia coast are almost entirely protected from the long swell of the open sea, which does, however, make itself felt in a few places, such as Dixon Entrance and Millbank Sound, and as a rule the water's surface is placid and unruffled to a remarkable degree, though it may be disturbed to a point of choppiness by local winds. This surface smoothness is deceiving, however, and underneath there is often a swiftness and strength to the surging currents that may well spell danger to any unaccustomed seaman, and that, in fact, brings a fairly regular return of accidents, year by year, to even those vessels that ply regularly along the coast.

The tundra or muskeg zone in southeastern Alaska is restricted in area and bears slight resemblance to the limitless tundra of more northern latitudes. It occurs merely as relatively small and isolated open areas in the surrounding forests, "parks" in the local parlance. There is a heavy ground covering of mosses and other low-growing plants which hold the moisture, and the entire area is well soaked, often partly submerged. It is noteworthy that the lodgepole pine (*Pinus contorta*) should in this region be found only in these bogs, and it seems to lead a most precarious existence here, at an extreme outpost of its vast habitat. In the interior at the same latitude (where it is known as "jack pine") it covers considerable areas, sometimes in pure stands over square miles of territory, but on the coast it is found in stunted form, growing as scattered trees over the wet "parks" that characterize some of the islands. It is not easy to understand how these pines managed to push westward as they have done, to occupy the scanty tracts suitable to their needs, where, in fact, they seem barely able to survive, but there they are, out clear to the westernmost edge of the Archipelago.

The Alpine-Arctic zone of the mountaintops includes the open, unforested areas that lie above the limit of upright trees, the timberline, which in this region is found at about 2,500 feet elevation or less. It is most extensive in area on the three large, northern islands, Baranof, Chichagof, and Admiralty, and is entirely absent from some of the low-lying southern islands. This area is, of course, cold and rigorous as to climate, snow covered over much of the year, and ice covered in many places. The vegetation consists of low-growing grasses, of heathers, lupines, stone-crop, and other small and hardy shrubs.

CASSIAR AND OMINECA DISTRICTS: TOPOGRAPHY, CLIMATE, AND PLANT COMMUNITIES OF THE INTERIOR

East of the separating Coast Range lies the vast interior of northern British Columbia and southern Yukon Territory, as different from the closely adjacent Sitkan district, it is no exaggeration to say, as Maine is from Florida. In topography, in climate, in fauna and flora, the contrasting differences are such as to strike even the most casual observer, it would seem, while to the close student of animal and plant life it is fascinating to see the replacement of species in the two areas. There is hardly a single kind of bird or mammal that is found commonly in both regions; for that matter there are but few species that occur at all in both.

In the region referred to, my observations have extended eastward to Hazelton and Kispiox Valley in the Skeena River drainage, to the Telegraph Creek region on the upper Stikine River, and to Teslin Lake, east of the Atlin region, each of these points being roughly 200 miles from the coast. The Stikine and Skeena rivers drain into the Pacific, and, in common with the other streams of the Pacific drainage, they serve as highways and spawning grounds for vast hordes of migrating salmon of several species. Only a short distance east of Telegraph Creek there are small streams flowing eastward that find their way to the Liard River and thence to the Mackenzie River and the Arctic Ocean. In the Atlin region most of the water (except the Taku, to the southwest) flows northward into the Yukon drainage, to empty into the Bering Sea. Streams forming the local sources of these three far-reaching watersheds interdigitate, approaching each other very closely, to within a few miles.

This is a country of broad valleys, of mountains (not nearly so rough and forbidding as the coastal barrier), and of magnificent lakes and rivers. The Coast Range presents on its eastern face nearly as abrupt a slope as on the coast, though it rises from a higher level (the valley floor at Atlin is about 2,200 feet altitude, at Telegraph Creek, 540 feet, and at Hazelton about 1,000 feet), and lesser spurs and separate scattered smaller ranges extend eastward in diminishing height and area, to the rolling plateau that stretches still farther east.

The climate of this interior region presents various sharp contrasts to that of the coast. East of the mountains there is abruptly a great falling off in amount of rainfall. This is especially true in the Carcross, Atlin, and Telegraph Creek regions, each with an annual precipitation of some twelve to fifteen inches, compared with one hundred inches or more on the nearby coast. The Skeena River flows to the sea through a broad valley that is bordered by mountains of lesser height than those to the northward, and coastal fog and rain drift inland here farther and heavier than is the case on the Stikine and Taku.

Inland the summers are clear and sunshiny, far warmer than on the coast, and the winters are extremely cold, again in contrast to coastal conditions, where winter weather is not so severe.

The interior is generally forested, as is the coast, but the inland woods, open, sunny and park-like, are a delightful contrast to the gloomy coastal jungles; the component trees and shrubbery are for the most part very different. The lowland forests of the interior are composed so largely of "poplar" (that is, the quaking aspen *Populus tremuloides*) that one may be pardoned a first impression that there is little else in the makeup of the woods. The bright green of the summer foliage and the brilliant yellows of the autumn are such conspicuous and forceful colors also, as to fix the attention almost subconsciously upon the dominant poplars, to the exclusion of the really abundant but less assertive conifers.

The poplar woods are pleasant to be in at any time, but in the fall, after the first frosts have come and while the brilliantly colored leaves are still on the trees, there is one peculiarly charming quality developed, one that is most manifest on gloomy, overcast days. The yellow foliage has exactly the tone of brilliant sunlight, and, stepping from the dense shade of spruce timber into a belt of poplars (the transition is usually abrupt), one receives the impression of a sudden clearing away of the clouds overhead, with the resultant burst of sunshine. Time and again under such circumstances I have had to correct an erroneous impression of this sort, a conviction so strong that it was hard for one to be otherwise convinced. Poplar woods in dense stands extend far to the northward of this section. They are characteristic of the lowlands of the Atlin region, and, to the southward, of the Telegraph Creek and Hazelton regions, and far beyond. Still farther south these trees climb to higher altitudes, throughout the Rocky Mountains, and (at altitudes of 8,000 feet and over) even into northern Mexico.

There are other deciduous trees of the interior that are of less general distribution. Balm of Gilead (*Populus balsamifera*) is common in the Atlin region and most easily picked out from the closely similar poplars in the early fall, for the Balm of Gilead is first to change color with the frost, and overnight, perhaps, may spring into yellow prominence. Birch trees are fairly common on the Stikine, sometimes of considerable size, but I saw none around Atlin. Huge cottonwoods are characteristic of the Stikine bottomlands. Far up river there are stately rows of them at Glenora, and also of the Skeena River and of the Kispiox. Willows are everywhere, from dense jungles in the river beds to absurd little fingerling growths on the bleak mountaintops, above almost all other vegetation, but at the most they are bushes, rarely attaining a tree-like aspect.

A characteristic timberline plant is the trailing birch (*Betula glandulosa*), common almost everywhere above the limit of upright trees. This covers vast

Thomas Bay, southeastern Alaska, 1909. Photograph by Harry S. Swarth, with permission of the Museum of Vertebrate Zoology, University of California, Berkeley.

areas, sometimes solid square miles, usually as a prostrate, trailing shrub that is easily walked over, then, in more favorable locations attaining to upright growth, knee-high, waist-high, or even higher. It thus forms what in California would be called "chaparral," in general appearance not unlike the scrub-oak covering the foothills of the southwestern mountains, but for the most part more easily traversed. This birch grows usually in disconnected patches, so that one can slip through or around, or over if need be, and, by picking and choosing and not being stiff-necked about immediate direction and right-of-way, generally wriggle through without too great difficulty. In the fall, when the poplars turn yellow, the birch flames into brilliant reds, and, together with fireweed and some smaller shrubs, paints the higher slopes in acres of gaudy colors.

Of the conifers, the conspicuous lowland tree is the white spruce (*Picea canadensis*), usually forming extensive tracts in almost pure stands where conditions are favorable. This is a tall, attenuated tree, with numerous short, downward pointing branches, and almost never with any long branch. With thickly clothed trunk, the whole tree is exaggeratedly slender, a characteristic feature that cannot be mistaken. Sometimes, not commonly, the crown becomes grotesquely enlarged, assuming curiously bulbous outlines, but as a rule the spruce forest thrusts an array of slender spear points upward, unbroken in general aspect for many miles.

In the upper Skeena Valley I saw the black spruce (*Picea mariana*), in small numbers and always in the same sort of places. The bits of marsh land

(muskegs) with which the generally dry forests are studded, are always en-
circled with a border of black spruce (often together with a few yellow cedar),
dark and funeral looking, with long streamers of pendant moss. This tree should
occur also in the Stikine and Atlin regions, but I did not see it in the sections I
visited. The Engelmann spruce (*Picea engelmannii*) also reaches the vicinity of
Hazelton, in the Skeena Valley, which is supposed to be its northern limit.

On sandy flats the jack pine is found, mostly in rather open groves, and in
a charmingly park-like manner of growth. In the spruce woods there is a tangle
of underbrush and debris on the ground that often makes traveling difficult,
but the pines are delightfully free of obstructions. True, there are places where
jack pines grow in abominably close array as to forbid travel by the very mass
of the crowded trunks, and I saw a number of such stands in the upper Skeena
Valley, but on the upper Stikine and about Lake Atlin, the groves were nearly
all open and park-like, as described.

The spruces occupy favored sections in the valleys and they cover thickly
the lower slopes of the mountains, but a little higher up they give way to the
alpine balsam (*Abies lasiocarpa*). This, the timberline tree of the region, attains
as goodly a height as the spruce along the middle altitudes (at from 3,000 to
3,500 feet in the Atlin country), but on higher slopes it becomes stunted, and
in many places utterly prostrate. Groves of gnarled and twisted balsam, growing in thick masses, with sprawling branches near the ground, their crowns not
more than thirty or forty feet high, are to be avoided as being among the most
difficult of all woods to traverse. On the rolling plateaus that form so many
mountain summits here, the nearly prostrate balsam thickets can be walked
over with ease, the thickly matted branches, sprawling some two or three feet
from the ground, supplying a sturdy support and reasonably secure footing.

The balsam is a tree with delightful personality and with strongly marked
characteristics that can impress even a non-botanical observer. A glimpse of
the growing cones is sufficient for identification of the tree, for, in contrast to
those of all the other conifers of the region, they stand bolt upright upon the
twigs. The cones are unsatisfactory things to keep, for as they dry the scales fall
off in utter disintegration, and the same thing happens upon the tree, where
the slender cores of the old cones remain standing, like so many tiny Christ-
mas candles. Then, in early summer the trees produce an amazing quantity
of yellow pollen, so that in clambering through the prostrate thickets, as a
naturalist is bound to be doing, one emerges covered with a saffron deposit,
as though thoroughly dusted with Colman's mustard. A secretion that is less
easily disposed of is found in the blisters of the semi-liquid resin upon the
trunks, which burst when pressed upon and smear their sticky contents upon
clothes and hands.

The last conifer of our brief northern list is the juniper (*Juniperus communis*), which can hardly be termed a tree. It grows in valleys and on summits, but always the same, a sprawling discouraged looking shrub, clinging closely to sheltering ledge or boulder, and usually as scattered plants.

The subarctic north is the country of berries, and especially so in the interior. Strawberries in open dry woods and in meadows, raspberries in the rock slides, mossberries in the damp places, are all delicacies to be accepted gratefully wherever found; there are plenty of others, such as the high-bush cranberry (whose sweetish odor hangs heavily in the air of the autumn woods), that serve well as preserves, and there are some, like the soapberry, that are only fully appreciated by the Indians and the birds. Many of the berry-bearing shrubs that grow close to the ground have fair portions of their crops covered by winter snows, to be kept in cold storage until spring-time, then to be utilized by the arriving birds. In the upper Skeena Valley, hazel bushes bear abundant crops of nuts, as is reflected in the name of the town of Hazelton.

Attractive as is all of the subarctic north, in its varied woods, lakes, and streams, to my notion there is none of it so beautiful, so compelling in its features, as the open country above timberline. As one goes southward the ascent from valley to summits becomes increasingly arduous, and in the Hazelton region there is a long, stiff climb through a broad intermediate belt of dark spruce covered mountainside, hard to traverse and very "dead" as regards animal life, before the timberline region is attained. At the latitude of Atlin, however, with the valley level at about 2,200 feet altitude and timberline at

Waterfront, Wrangell, Alaska, 1924. Photograph by Harry S. Swarth, with permission of the Museum of Vertebrate Zoology, University of California, Berkeley.

about 3,500 feet, the climb is relatively inconsequential; about an hour's hard work takes one through the forested belt.

On the summits there are frequently many square miles of nearly level ground, rolling plateaus or terraced steps that are approached by gentle slopes. The groves of balsam at the general timberline border are extended in occasional far-reaching tongues onto the lower summits, or into sheltered high mountain valleys. There, lines of trees, or more often small isolated groves, give an attractive, park-like aspect to the landscape, heightened by the banks of bright flowers that are characteristic of midsummer conditions. Small, shallow lakelets, streams running between mossy banks or between overhanging masses of dwarf willow, melting snowbanks, wasting away as the summer advances, all in open country that can be walked over with ease almost everywhere, form an irresistible combination of attractions. This all refers to elevations (in the Atlin region) below 5,000 feet. Above that level the ridges are apt to be gravelly, relatively bare, though with a fairly thick covering of grass, and with a greater proportionate area of rocky slopes, rock slides, and piled up boulders, larger and more nearly permanent snowbanks, and a generally harsher and bleaker aspect.

LIFE ZONES AND FAUNAL AREAS

In the study of the distribution of North American birds, mammals, and plants, various concepts have been used. Early impressions as of "eastern," "central," and "western" regions were gradually abandoned as unsatisfactory, and even thorough and accurate studies of various different areas remained for a time uncorrelated by any general principles. Then Dr. C. Hart Merriam (at the time Chief of the Biological Survey of the United States Department of Agriculture) began publication of a series of studies of regions in western North America that so satisfactorily substantiated his theory of life zones. See especially, *Results of a Biological Survey of the San Francisco Mountain Region and Desert of the Little Colorado, Arizona*, by C. Hart Merriam, United States Department of Agriculture, North American Fauna no. 3, 1890; and *Results of a Biological Survey of Mount Shasta, California*, by C. Hart Merriam, United States Department of Agriculture, North American Fauna no. 16, 1899.

This theory recognizes but two primary life provinces in North America, a northern (Boreal) and a southern (Sonoran), the origin of all forms of life being claimed to have been from one direction or the other. These provinces are subdivided, the Sonoran (its eastern extension termed Austral) into Lower (southern) and Upper (northern); the Boreal, from south to north, into Canadian, Hudsonian, and Arctic (the Arctic of the southern mountaintops called

Alpine-Arctic); and between Sonoran and Boreal there is inserted the Transition Zone, neutral ground with a mixture of species from both sources. All of these belts, with a general east and west trend, are tremendously affected by altitude, so that in mountainous western North America there is a most complicated interdigitation of the Sonoran zones northward in hot lowlands, of the Boreal Zones far southward at various mountain levels, the altitudinal feature being further complicated by slope exposure (south facing slopes being warmer, hence "lower" zonally, north facing slopes colder and higher, zonally) and some other minor factors. The primary divisions are severally occupied by distinct genera, the lesser divisions by distinctive species. All the divisions are based mainly upon temperature, especially maximum summer temperatures. An enormous literature on the subject has grown up, and there are masses of published data available for many sections of the country that enable anyone to test for himself the application of the theory.

Later on another concept, of "faunal areas," was employed by Joseph Grinnell, based largely upon relative humidity, and, in North America, most satisfactorily exhibited in the west, with its varied topography. A faunal area, perhaps less than one hundred miles square, may comprise several life zones, and a life zone in its various extensions across the continent will traverse a number of faunal areas.

In the northwestern section of which we treat here, the region west of the Coast Range in southeastern Alaska, the Sitkan district, is a well-defined faunal area, too far north for Sonoran, Transition, and Canadian zones, and at different elevations comprising the Hudsonian Zone at the lower levels, Alpine-Arctic on the summits. East of the Coast Range is the Cassiar district, another faunal area, Hudsonian zone over most of the lowlands, and Alpine-Arctic above timberline.

Apparently the Cassiar district should be limited southward along an east and west line about one hundred miles north of the Skeena River at Hazelton. There is a rise of land thereabout that marks the northern limits of a number of bird and mammal species, as well as reptiles (only one species, a snake, *Thamnophis*, extends that far north), and there are many northern birds and mammals that have their southern boundary about at the same line. With the southern delimitation of the Cassiar district tentatively thus placed, there remains the further definition of the faunal area that lies south of that line. For this we lack sufficient data, other than that the Coast Range will still serve at the west.

Altogether, piecing together our available facts regarding the occurrence of birds, mammals, reptiles, and plants in northern British Columbia and southeastern Alaska, we can reach the following conclusions. That the Sitkan district can be recognized as a well-defined faunal area, with Lynn Canal and Glacier Bay at the northern extremity, Dixon Entrance at the south, and including all the islands between those extremes, as well the extremely narrow ribbon of the adjacent mainland coast, westward from the summit of the Coast Range. The Cassiar district can be partly outlined, its western boundary the Coast Range, its northern limit arbitrarily placed for the present along the Yukon–British Columbia boundary line, its southern boundary at the rise of land about one hundred miles north of the Skeena River at Hazelton, its easternmost extension for the present undetermined. South of the Cassiar district and east of the Coast Range the existence of another distinguishable faunal area is suggested, which may be called the Omineca district (after the mining district, so-named, of the same general region). The southern and eastern boundaries of this faunal area are still to be determined.

The narrow strip of territory lying west of the Coast Range and the Cascades possesses strongly marked characteristics of fauna and flora that stamp this whole coastal area from Alaska south into California, in some respects even as far as southern California. The Sitkan district is almost the northern extremity of this peculiar coastal strip. Of the birds of the Sitkan district, with which we are here mainly concerned, with but few exceptions they are the same as, or most closely related to, species farther south along the coast, and for the most part far more remotely related to those of the closely adjacent interior. Characteristic coastal birds that at once come to mind are the Sooty Grouse, Dusky Horned Owl, Red-breasted Sapsucker, Red-shafted Flicker, Western Flycatcher, Steller Jay, Beach Crow, Oregon Junco, Rusty Song Sparrow, Lutescent Warbler, Chestnut-backed Chickadee, and Russet-backed Thrush. On the northern islands and on the Alpine-Arctic heights there are some boreal species, such as the ptarmigan and Rosy finches, but these are not many, and there are many more boreal birds that extend to at least as far southern latitudes in the interior and that eschew the coast altogether.

The seasonal migration of species to and from this region is not the noticeable feature that bird migration is in most parts of North America. There are a few birds that do have well defined and fairly extensive migrations, such as the several species of swallows, Western Flycatcher, Lutescent Warbler, and Russet-backed Thrush, which withdraw in winter as far as Mexico and perhaps Central America, but these are exceptions. Mallards and White-cheeked Geese remain through the winter, as well as some other ducks and many seabirds. Some small birds that do migrate travel but a short distance southward, perhaps to the coast of southern British Columbia, or less commonly, into California. Some Song Sparrows, chickadees, Steller Jays, and even occasional Townsend's Fox Sparrows, American Robins, and Varied Thrushes, find

Expedition boat (left) at Juneau, Alaska, 1909. Photograph by Harry S. Swarth, with permission of the Museum of Vertebrate Zoology, University of California, Berkeley.

favored sections of the Sitkan district where they stay throughout the winter months.

It seems appropriate to speak here of a few mammal species as offering distributional facts to what is seen in the birds. The three large northernmost islands, Baranof, Chichagof, and Admiralty, differ from the southern ones in several respects, notably in the occurrence there of species of the huge northern brown bear, to the exclusion of the black bear, which occupies all of the southern islands. The characteristic ungulate of the Sitkan district is the black-tailed deer, common on all the islands, rare on the mainland; inland, the moose occupies a corresponding position. Caribou, mountain sheep, lynx, fox and rabbit are characteristic northern mammals that are common inland, almost unknown to the coastal district. This statement must be modified to this extent, that a form of caribou has been found upon one of the Queen Charlotte Islands (south of the Sitkan district). This was a most unlooked-for occurrence, and the animals were apparently few in numbers and of dwarfed stature, and the species may even now be extinct. The common meadow mouse of the southern and central islands of the Sitkan district is *Microtus macrurus*, which ranges south at least as far as Puget Sound. The most common inland species is *Microtus pennsylvanicus drummondii*, the western representative of a species that extends eastward across the continent. *M. p. drummondii* occurs commonly from the interior along the Taku River clear to saltwater, the only place where it is known to reach the coast. And on Admiralty Island, directly across the channel from the mouth of the Taku (and on that island alone), there is a meadow mouse that is barely (dubiously, perhaps) to be distinguished from *drummondii*. It is a suggestive occurrence. Then, coastwise and inland, there are in the two regions related, but distinct, species or subspecies of red squirrel, weasel, marten, white-footed mouse, red-backed mouse, and others, the conditions (allowing for our less exact knowledge of the small mammals) being about as with the birds. Among the mammals, just as among birds, there are in the Sitkan district more species along the mainland coast and on the innermost islands, and a progressive diminution in number of species westward to the outermost edge of the archipelago.

There are some notable peculiarities in the distribution of birds and mammals among the islands of the Sitkan district. We are accustomed in western North America to find the intangible barriers formed by differing degrees of temperature and humidity acting with potency and exactness, but in this region such factors are practically uniform throughout. There are, however, the mechanical barriers formed by the channels between the islands, barriers whose effectiveness we might be expected to gauge pretty accurately, in relation to their depth, width, and the character of the species involved. As a matter of fact, we find in many cases a disconcerting lack of conformity between the range of a species and the factors above listed that might be expected to control its distribution. Certain species of grouse and of mice may be considered together in this connection.

Coronation and Warren are two small outlying islands at the western edge of the Alexander Archipelago. Coronation lies about six miles south of the large Kuiu Island, Warren about two miles west of Kosciusko (which is barely separated from Prince of Wales Island). Coronation and Warren are about eight miles apart. They have one species of meadow mouse (*Microtus coronarius*) common to the two, and as far as known occurring nowhere else; and a white-footed mouse (*Peromyscus sitkensis*) found upon both which is not found on either of the large islands closest to them. Now, the Sooty Grouse (*Dendragapus*), resident on Kuiu, is found on Coronation, but not on Warren. The Spruce Grouse (*Canachites*), resident on Prince of Wales, is found on Warren but not on Coronation. It is hard to imagine what the nature of the factors may be that have permitted small and feeble terrestrial mammals (mice) to establish themselves upon two islands, eight miles apart, but have prohibited their further dispersal over the lesser channels that separate them from the larger islands beyond: and at the same time have barred strong flying species of grouse which have already crossed the narrower channels. It may be said further, that elsewhere in the Alexander Archipelago both species of grouse are delimited in their distribution by channels four miles or less across, while both must at some time have passed over far wider channels in order to have crossed from the mainland to the islands where they now are. Nowhere are both species known to be upon the same island. They have with little doubt been established upon the islands for vast periods of time, but it is hardly conceivable that this territory was free of ice and habitable for animal life until long after the separation of islands from mainland. There is much evidence tending to show the gradual extension of bird and mammal species westward from the mainland to islands that were already in existence as such. Of both birds and mammals, the existence of species on the adjacent mainland that are on none of the islands, the greater number of species on the islands near the coast, and the steady lessening in kinds and individuals westward; also, the known distribution of certain forms over groups of islands obviously reached with ease, one from another; all these are facts supporting the theory mentioned. At the same time there are more than a few curious, apparently anomalous, occurrences and non-occurrences upon certain islands that are puzzling in the extreme.

Bohemian Waxwing nest with eggs, Telegraph Creek, British Columbia, 22 June 1919.
Photograph by Joseph Dixon, with permission of the Museum of Vertebrate Zoology, University of California, Berkeley.

Of the avifauna of the Cassiar district, in strong contrast with that of the coast, affinities are all toward boreal and eastern North America. There is a strong representation of northern species, many of them resident, and of eastern birds, most of which perform prodigiously long migrations. Of the former there are such species as ptarmigan, Gyrfalcon, Hawk Owl, Canada Jay, Rusty Blackbird, White-winged Crossbill, Redpoll, Bohemian Waxwing, Northern Shrike, and Hudsonian Chickadee. Of eastern birds, some characteristic species are Yellow-shafted Flicker, Eastern Nighthawk, Eastern Purple Finch, Tennessee Warbler, Eastern Yellow Warbler, Myrtle Warbler, Blackpoll Warbler, Redstart, Olive-backed Thrush, and Eastern Robin. There is, besides, an infusion of Rocky Mountain and Great Basin species, such as Say's Phoebe, Bendire's Crossbill, Townsend Solitaire, and Mountain Bluebird.

In the Omineca district we are south of the line where truly boreal species may be found nesting in the lowlands, and such birds as Hudsonian Chickadee, Canada Jay, and White-winged Crossbill must be sought on the higher mountain slopes, to which the Hudsonian Zone is there restricted. Here, in the valleys, we find a number of birds at their extreme northern limits, some of them again being eastern species, such as the Eastern Kingbird, White-throated Sparrow, Red-eyed Vireo, Magnolia Warbler, and Catbird. The Audubon Warbler is common here, just south of the southern limit of the closely related Myrtle Warbler, and here, too, is approximately the northern bounds of Pileated Woodpecker, Western Crow, Evening Grosbeak, Cedar Waxwing, and Western House Wren.

The Coast Range in southeastern Alaska forms a barrier between coastal and interior regions of an abruptness and effectiveness that can hardly be duplicated elsewhere among North American mountain chains. It rises steeply from saltwater in a series of serrated ridges and towering peaks that culminate in summits 8,000 feet high and more. The highest elevations appear to be jagged and exposed masses of granite, below which glaciers and snowfields cover most of the middle heights and fill many of the lower canyons and valleys. Unbroken forest clothes the lowest slopes. It has been said that no mountain chain, however rugged and steep, is itself a barrier to animal distribution, since any mouse (that is, the species) would be able eventually to cross, but that the climatic changes, of temperature especially, that accompany rising altitudes are what really constitute the obstacle. This is, of course, true enough, even in final analysis, of the case here considered, but it would seem that towering granite cliffs, and many miles of glacial ice, would in themselves discourage passage of animal life, even of kinds that could otherwise endure such conditions.

In traversing this barrier, as we readily may over the summit at White Pass, or piercing the center along the courses of the Stikine and Skeena rivers, it is easily possible to recognize the obstructing conditions that exist throughout this belt. In ascending the White Pass, from coastal forest conditions at the sea level of Skagway (not quite typical of the coast region generally, perhaps, but sufficiently so), one arrives rather abruptly at the timberline surroundings of the summits at "The Boundary." The long stretch of tundra-like country that is then traversed, all at high altitudes, is obviously unsuited to the forest dwellers at the coast, both birds and mammals, most of which never very nearly approach the upper timber limit. In descending the eastern slopes of the White Pass a somewhat more gradual entry is made into the forested lowlands, the dwellers of which are as rigorously confined to their side of the mountains.

The Stikine River rises far in the interior, flows westward through the Coast Range and reaches saltwater near the town of Wrangell. In its passage through the mountains, a distance, roughly, of about one hundred miles, the river valley is much constricted, hemmed in between steep mountainsides, sometimes the river itself confined in narrow gorges between perpendicular walls of rock. Quotations may be taken from a writer upon the Tahltan Indians, who thus describes this section of their country:

> The lower valley of the Stikine from just below Glenora to the coast, a direct distance of about eighty miles, is included within the coastal range and constitutes a region of great humidity, with leaden skies and an annual precipitation equaling if not exceeding that of the coast which reaches a mean of eighty-six inches. The snow thereabouts is excessive, and accounts for the extensive glaciers that fill the valleys; and long after spring has opened in the colder interior the lower river flats are covered with their burden of snow and ice. . . . Forests of spruce, fir, cedar, and hemlock cover the mountain slopes to the limit of tree growth, while in the river valleys cottonwoods grow to considerable size, and groves of alder and willow, with the devil's club and berry bushes, form an almost impenetrable barrier. . . . It may be pertinent to remark here, that this region which may be characterized as the wet belt has never been inhabited by either Tahltan or Tlingit in the sense they have permanently occupied it and it is scarcely more popular as a hunting ground owing to its poverty and inaccessibility. Emmons, G. T., The Tahltan Indians, University of Pennsylvania, The Museum, Anthropological Publications, vol. 4, no. 1, 1911.

Thus, the valley of the Stikine, which might be regarded as a potential highway between coast and interior, actually does not act as such, so far as animal and plant life is concerned, or at most to a negligible extent. The uniformity and extraordinary density of the all-prevailing forest would debar the presence of all but a few animal species in any event, and the persistence of winter conditions long after spring has opened upon either side of the mountains is a climatic barrier that is doubtless of strong potency. There are most abrupt changes in fauna and flora on the two sides of the Coast Range where the Stikine pierces the range, and only a slight overlapping of species along this river valley.

The Skeena River, like the Stikine, has its sources far inland and reaches the coast through an opening between the mountains. The Coast Range at that point, however, is not so lofty and rugged as it is farther north, and the Skeena flows through a broader valley than does the Stikine. Coastal rain and fog drift inland here to a much greater extent, and there is also along the upper Skeena (the northern Omineca district) an intermingling of coastal and inland birds, and of some forest growths, such as cannot be found in the Cassiar district. This is exemplified in the character of certain bird species, such as the Steller Jay, the junco, and most conspicuously, in the two flicker species. Here, too, the Tlingit Indians of the coast have penetrated far inland, and there are old established villages with rows of ancient totem poles in the valleys of the Skeena and its tributary the Kispiox, one hundred miles from the coast.

Thus, observations of conditions along various lines between the coast and the interior, as the White Pass, Stikine Valley, and Skeena Valley, yield many significant facts bearing upon the distribution of species in the two regions. Certain supplementary observations that are available, from field work at the heads of certain far-reaching inlets and at the mouths of lesser streams that also rise east of the mountains, all support the same general conclusions.

AVIFAUNA OF THE SITKAN DISTRICT: COASTAL REGION

Any speculations on the origin of the avifauna that now occupies the Sitkan district would seem to be safely based on an assumption of the relatively recent occupancy of the territory by animal and plant life of any sort. Confirmation of the earth's earlier surface, the vast areas that are ice-covered at the present time, and the fact that these ice fields are visibly retreating today, all point to the likelihood of this region having been ice-bound in its entirety at a not remote period, speaking in terms of geological time, perhaps not so very remote even in terms of years.

With the bird population in mind, it seems likely that the country east of the mountains became clear of ice and habitable long before the coastal district, to be occupied at once by the species to the southward and eastward.

The Coast Range stood as an impassable barrier to this immigration, marking the bounds of a long array of species, as it does today, even after the coastal region eventually became habitable in part to the quite different array of species that flowed in from the country directly to the southward. The presence of this population would, presumably, exert against the eastern species pressure of another sort, of an effectiveness to be tested after the eventual nullification of the mechanical barrier found in the ice-clad mountains. There can be seen today in places the beginnings of this passive struggle between unlike species with similar needs, and between closely related species now meeting after traveling widely separated avenues of approach.

At least, that is the way it looks to me. There is no appeal to me in the alternative view that the whole region, east and west of the Coast Range, might have been originally occupied by an ancestral population that was all alike, to be variously modified subsequently by the varied conditions of the several sections.

Truly Arctic birds, which form but a small proportion of the avifauna of the Sitkan district, were probably the latest arrivals there, for the timberline habitat to which most of them are restricted would be about the last part of the country to become available for occupancy.

The Sitkan district is poor in the number of landbird species that comprise its avifauna, and for over the greater part of the region there is also a notable paucity of individuals. At certain favored places, and at some periods, birds can indeed be found in numbers, but there are hundreds of square miles of unvaried spruce forest, dark and cheerless, and almost birdless. On some of the westernmost islands in particular one can easily yield to an impression that everything is so new and unfinished that the birds have not yet found the place. Number of species and number of individuals lessen steadily from the mainland coast westward. A reason for the relatively few species on most of the outer islands lies in the uniformity of environment, and of an environment that is not adapted to a variety of bird life. Unbroken coniferous forest, densely massed, prevails over three-fourths of the area, this bordered along the beaches by a narrow belt of alders and surmounted by a great or less expanse of Alpine-Arctic summits. Small acreages of salt-grass meadow on the tidal flats, of muskegs or parks on the hill-sides, of densely brush-covered streamsides and lake shore, and of alder-grown slides on the steeper slopes, comprise about as much variety as can be found.

Along the mainland shore there are the mouths of several large rivers that pierce the Coast Range, and in these broad valleys there is to be found a greater variety of surroundings, and an environment, too, that is much more favorable to a long list of bird species. These river valleys also form avenues of approach from the country beyond the mountains, and exploration at the mouths of such streams has disclosed the presence in small numbers of inland bird species not occurring elsewhere in the Sitkan district. Collecting by the writer one summer along the lower reaches of the Chickamin River, some seventy-five miles north of Dixon Entrance, yielded specimens of Violet-green Swallow, Cedar Waxwing, Louisiana Tanager, and Alder Flycatcher, all inland species that had not then been found elsewhere upon this Alaska coast, but which were certainly breeding at that point. There, and at such similarly situated places as the mouths of the Stikine and Taku rivers, and at the heads of some of the deeper inlets, other inland birds occur in limited numbers such as Olive-sided Flycatcher, Western Wood Pewee, Western Yellowthroat, and Tolmie Warbler.

These advancing scouts may fairly be considered as the worthy descendants of hardy ancestors who ages ago extended the domain of their kind as rapidly as the retreating ice left territory open for occupancy. They are entering a region now where some will find limited areas suitable to their needs unoccupied by any bird species; some may find a cousin with like tastes already on the ground to dispute a later arrival. Franklin's Grouse is an inland species that has succeeded in entrenching itself on several of the coastal islands, so recently that little or no change in appearance of the bird is yet apparent. Its northern cousin, the Valdez Spruce Grouse, has also reached one or two coastal points, and has developed a barely perceptible color change from the parent stock of the interior. Hybrid flickers that have been found on the coast and inland give evidence of the meeting of the Yellow-shafted and Red-shafted species, with the same results here as elsewhere under similar conditions. The Alaska Three-toed Woodpecker and the Rocky Mountain Downy Woodpecker have both reached the Sitkan district in scanty numbers, the former established as a resident, the latter known only as a migrant, neither of them perceptibly altered in color or markings. The Hairy Woodpecker of the Sitkan district reached that region from the interior, a slightly modified variant of the Rocky Mountain Hairy Woodpecker, and is firmly established there, of general distribution and of fair abundance, as birds go in that country. This subspecies is not particularly closely related to the dark-colored Harris Woodpecker of the southern British Columbian coast, with which it was formerly considered to be identical. That form has developed a still darker variety upon the Queen Charlotte Islands, the northern limit of this dusky strain. The avifauna of the Sitkan district is mostly derived from the coast to the southward, but there are some elements, and the Hairy Woodpecker is one of them, that unquestionably had their origin in the hinterland to the eastward.

Lake Atlin and town, 1924. Photograph by Harry S. Swarth, with permission of the Museum of Vertebrate Zoology, University of California, Berkeley.

The big rivers above mentioned form migration paths or rather by-paths, to a limited extent and used mostly by inexperienced young birds on their first southward flight. Magpies appear pretty regularly on the coast in the fall, and Mountain Bluebirds, Say's Phoebes, Sparrow Hawks, and even an occasional Northern Shrike and Rusty Blackbird drift downstream to saltwater at the summer's close. The Myrtle Warbler is a regular and fairly common migrant on the mainland coast. Presumably a slight modification of existing conditions might enable it to extend to the Sitkan forests the boundaries of its nearby summer home. It breeds now within eighty or ninety miles of the coast, and there is no obviously competing species in that region.

In contrast to the many eagerly coastward advancing pioneers from the east, pressing coastward wherever a possible opening appears, most of the components of the coastal avifauna conservatively cling to the harsh and exacting surroundings. A fair proportion of the coastal birds are largely or entirely resident there, and any range extension eastward would entail wintering under widely different (more severely cold) conditions than in their present home. So, with a few exceptions, the Sitkan birds remain pretty constantly on their own side of the mountains. One of the exceptions is the Rusty Song Sparrow, a local variety of a widespread species, a species that, most exacting in its requirement of fluviatile or littoral surroundings, is at the same time tolerant of an extreme variety of climatic conditions. From central British Columbia northward we can see the gradually increasing unfitness of the inland climate to this bird, for, extending across the province in its southern half, northward the song sparrow area becomes gradually pinched until it becomes a narrow ribbon along the coast. The Rusty Song Sparrow is abundant up the Skeena River at least to the Hazelton region, it is rare up the Stikine as far as Telegraph Creek, and the next step to the northward. A single bird that I once saw at the southwestern corner of Tagish Lake (just east of the Coast Range, in the Atlin region) was probably a straggler such as reaches that point from time to time. So, the Rusty Song Sparrow is feeling its way inland into a region still unoccupied by its kind, though to our eyes the country does not seem unfitted for summer residence of the species.

The Red-breasted Sapsucker is another bird that extends far eastward in southern British Columbia, but is confined to the coast in the north. Like the Song Sparrow, and to about the same degree, it presses far inland up the Skeena Valley, in much lesser force up the Stikine.

The northwest coast humid belt is commonly spoken of as being inhab-ited by dark-colored forms of birds and mammals, the rich browns distinctive of certain characteristic species of the region being taken to be representative of the mode. This is true to a certain extent, but it is a statement that requires considerable modification. For one thing, as regards birds, the darkest-colored varieties occur mostly south of the Sitkan district, on the Queen Charlotte Islands and in the Vancouver Island and Puget Sound regions. This is apparent in such variable forms as the Steller Jay, Hairy and Downy woodpeckers, Song Sparrow, and Fox Sparrow. Besides the large proportion of resident landbirds in the Sitkan district there is a fair number of summer visitants, and these are mostly species that breed all along the coast south to California, and that spend the winter still farther south. Of these species the Sitkan contingent is for the most part not changed one whit from the mode of their kind, and Alaska specimens of, for example, Western Flycatcher, Lutescent Warbler, and Russet-backed Thrush (all rigidly confined to the coast in the north), are not to be distinguished from the same species in California. There are others that are more widespread in their general range, such as Barn Swallow, Tree Swallow, Belted Kingfisher, and Water Ouzel, all fairly common in parts of the Sitkan district, that again have not developed any peculiarities of color in that part of their habitat.

It is among the resident or nearly resident species, those that at the most migrate but a short distance up and down the coast, always within the same general type of environment, that we find the rich, "saturated" coloration that we have grown accustomed to associate with the humid coast region. In this category there are the locally named races of Great Blue Heron, Horned Owl,

Red-shafted Flicker, Steller Jay, Song Sparrow, Fox Sparrow, Ruby-crowned Kinglet, and Hermit Thrush, to name some of the more conspicuous and better-known examples.

Among the waterbirds the Canada Goose is the one outstanding species that has developed a strongly characterized local race on the northwest coast. This subspecies, the dark-colored White-cheeked Goose, is closely confined to the coastal region, is almost exclusively littoral in its choice of environment, and is very slightly migratory. The bulk of the population remains in parts of the Sitkan district throughout the winter. In this connection comparison with the Mallard at once comes to mind. The Mallard in this region has much the same habits, mode of life, and choice of environment as the White-cheeked Goose, and it is, similarly, resident in the Sitkan district the year through. While it has not been proved, it seems likely that like the goose, the species is resident in the sense that it is the same population that does remain there, not a moving out of summer visitants, replaced by winter visitants of the same species from other points. At any rate, the Mallard in the Sitkan district exhibits no perceptible peculiarities of appearance. The Canada Goose, however, is of a "plastic" and impressionable nature, in that it has developed several strongly characterized subspecies in different parts of its wide habitat across North America, while the Mallard, at home throughout the Northern Hemisphere, is notably resistant to local influences and remains unchanged over nearly all its range.

Any study of the birds of the Sitkan region should take into consideration the relation of this section to the Alaska Peninsula and the Aleutian Islands, which lie far to the westward and in part south of the latitude of the Sitkan district. There are various bird species known to inhabit those western portions of Alaska, that have been inferentially ascribed to the Sitkan district, too. Peale's Falcon, for example, because it nests also on the Queen Charlotte Islands, south of the Sitkan district, and several species of auklets because they appear in winter along the Washington coast, still farther south. Actually, the Sitkan district is not on a direct line between the Queen Charlottes and the Aleutians, nor between the coast of Washington and the Aleutians. Certain observations tend to show that there is a line of migration from the Alaska Peninsula to the southeast, almost entirely passing by the Sitkan district as well as much of the coast of British Columbia. The Aleutian Savannah Sparrow (*Passerculus sandwichensis sandwichensis*) is a case in point. This bird breeds on Unalaska and on the Aleutian Peninsula, and in migration it occurs sparingly on the western islands of the Alexander Archipelago. It is extremely rare on the inner islands, and it is, I believe, unknown along the Alaska coast immediately north and west of the Sitkan district. Apparently, the main migration route must lie from the Alaska Peninsula across the Gulf of Alaska, a little south of east, to the Queen Charlotte Islands and the coast of southern British Columbia.

AVIFAUNA OF THE CASSIAR AND OMINECA DISTRICTS: INTERIOR REGION

As regards the avifauna of the northwestern interior, of the Cassiar and Omineca districts, it is distinctly similar to that of eastern North America. Of northern species in its composition, of course, and in some cases with modified representatives (subspecies), as local exponents of the species, but still, on the whole, with little that is suggestive of the United States west of the Rockies, and almost nothing suggestive of the extreme western Pacific slope. The great stream of migrating birds that traverses the length of the Mississippi Valley washes against the eastern face of the Rocky Mountains, and that range trending more to the westward at its northern end, and at the same time becoming of less formidable continuity as a barrier, the more northern migrants press westward through passes over the lesser heights and occupy the lowlands beyond, clear to the Coast Range. The general tone of the summer bird population is given by such conspicuous and typical eastern forms as the Yellow-shafted Flicker, Eastern Nighthawk, Eastern Purple Finch, Blackpoll Warbler, Redstart, and Olive-backed Thrush, all of which reach extreme northern British Columbia, most of them considerably farther north. In the Skeena Valley, of the Omineca district, all of the above occur, and there are also many other eastern birds, here at their northwestern limit, such as Eastern Kingbird, Red-eyed Vireo, Magnolia Warbler, and Catbird.

Throughout both the Omineca and the Cassiar district there is a lesser representation of birds that occur in the United States from the Rocky Mountains westward. For the most part it has not yet been possible to recognize in the northern summer visitants any peculiarities that would definitely attach them to any particular southern region, but my conviction is that in more cases the individuals that summer in the north are related to the Rocky Mountain and Great Basin section of the species' range, rather than to the contingent of the Pacific slope. This group includes Swainson's Hawk, Say's Phoebe, Hammond's and Wright's flycatchers, Western Wood Pewee, Western Savannah Sparrow, Louisiana Tanager, Audubon's Warbler, and Rocky Mountain Bluebird.

Of Pacific coast birds that spend the summer in the northern interior there are Gambel's Sparrow and Golden-crowned Sparrow. The latter is the only species I know of that breeds inland in the north and winters in the relatively restricted Pacific slope region to the southward, the winter home of so many birds of the Sitkan district. The Golden-crowned Sparrow nests at higher altitudes in the Cassiar district and it is just possible that it does so in some parts of the Sitkan district as well, perhaps along the mainland coast. The mountain summits of the coast have not been thoroughly explored, and the Golden-crowned Sparrow is known to reach the coast of Alaska farther to the north and west.

There are certain absentees from the Atlin region that might be expected to be there, and their absence from this particular section supplies subject matter for speculation. This list includes Rosy Finch, Fox Sparrow, Pine Grosbeak, and Varied Thrush, each of which species there is a more northern, eastern, or southern subspecies, on the one hand, a more western subspecies on the other. It is a striking fact that instead of a population of connecting "intermediates," typical of neither one form nor the other, such as might have been looked for in the region, the several species are not represented there at all. It may be noted, too, that there are certain northern subspecies of the Hairy Woodpecker and Downy Woodpecker, for example, that occur in typical form but in very small numbers, as far south as Atlin, while a short distance farther south the southern subspecies of these species, again in nearly typical form, suddenly occur in some abundance.

We thus find in this northwestern country that there are certain slightly differentiated subspecies whose respective habitats are separated (in part, at least) by a hiatus in which neither form of the species occurs, and that there are other such closely related races which occur each in small numbers but in typical form almost or quite to a meeting point. In neither case is there the "intergrading" at these particular points of junction that we regard as one test of a subspecies. The observed conditions would seem to be a result of the manner in which the country was populated by the several subspecies. It suggests a filling up of available territory by forms with characters already fixed at the time of immigration.

There is a surprisingly long list of birds that breed in the Atlin region, but with extreme rarity. Of these, the Gyrfalcon (*Falco rusticolus*) and Northern Shrike (*Lanius borealis*) are probably rare and are at the extreme southern limit of their breeding range. So, too, at their extreme northern limit, are the similarly scarce Eastern Purple Finch (*Carpodacus p. purpureus*), Western Yellowthroat (*Geothlypis trichas occidentalis*), Western Warbling Vireo (*Vireo gilvus swainsoni*), and Mountain Chickadee (*Penthestes gambeli abbreviates*).

ACCOUNTS OF THE VARIOUS SPECIES UNDER CONSIDERATION

Loons, Grebes, Gulls, Terns, Auks, Storm-Petrels, and Cormorant

In the nearly absolute restriction of seabirds to the sea and their absence from freshwater, there is no question of life zones or faunal areas as explanatory of their distribution. True, ocean birds are confined within habitats that are circumscribed by as intangible barriers as those of life zones, but that is a phase of the matter with which we are not here concerned. It is to all appearance the love for saltwater or else the dependence upon some condition in, or production of, saltwater, that keeps seabirds in the sea and away from closely adjacent bodies of freshwater. I do not know whether murrelets, auklets, and their ilk would soon pine away and die if kept captive in freshwater, as some of the Antarctic penguins have been known to do, but normally these birds never appear in freshwater streams and lakes. So, from the waters of the Sitkan district there is a long list of seabirds that are unknown inland. There are also, however, some curious occurrences inland of bird species that are commonly regarded as saltwater habitants purely.

The northern grebes are the large Holboell's Grebe and the small Horned Grebe. Both seek freshwater during the nesting season, and they may be found side by side on most British Columbia lakes. Both (the Horned Grebe in far greater numbers) pass up and down the coast in spring and fall; they migrate commonly on freshwater or salt.

All the loons are abundant on saltwater during migration, and, farther south, throughout the winter. The Great Northern Diver, here near its northern limit, is a freshwater bird in the nesting season, common on British Columbia lakes, and probably nesting in lesser numbers on freshwater lakes on the Alaska islands. The Pacific Loon and Red-throated Loon are more northern species, the former especially so, and neither is averse to saltwater. The Red-throated Loon is known to breed on the coast as far south as the Queen Charlotte Islands. Incidentally, it may be pointed out that the Great Northern Diver, despite this label, is the most southern of his family, but he is by no means the only bird that is made to bear an utterly inappropriate name, one, however, that long usage has made to seem entirely proper.

The list of the Sitkan district diving birds, seabirds that are particularly and distinctly of the saltwater, includes Tufted Puffin, Horned Puffin, Rhinoceros Auklet, Cassin's Auklet, Ancient Murrelet, Marbled Murrelet, Pigeon Guillemot, and California Murre. Every one of these birds has a personal history that is well worth following up, but it may suffice here to call attention to the most common one of the lot, the Marbled Murrelet, which may be seen

Waterbirds.

Coast	Interior
	Holboell's Grebe
	Horned Grebe
Great Northern Diver	Great Northern Diver
Red-throated Loon	
Tufted Puffin	
Horned Puffin	
Rhinoceros Auklet	
Cassin's Auklet	
Ancient Murrelet	
Marbled Murrelet	
Pigeon Guillemot	
Common Murre	
Glaucous-winged Gull	
	Herring Gull
	Short-billed Gull
	Bonaparte's Gull
Arctic Tern	Arctic Tern
Beal's Petrel	
Fork-tailed Petrel	
Pelagic Cormorant	

anywhere and everywhere in the sheltered waters of the inner channels, and which despite its abundance, to this day defies the efforts of the egg collector. It is one of the very few species of North American birds whose nesting habits are practically unknown. Marbled Murrelets may be seen from the steamers almost any day between Seattle and Skagway. When in the vessel's path they flap painfully out of the way, wings used as flippers, with an appearance of crawling upon the surface of the water, and conveying an impression of ut-ter helplessness. If approached too closely, however, they dive, and they can dive, swim, and fly exceedingly well, despite their incompetent appearance. The diving birds listed above have no prototypes in the freshwater streams and lakes of the interior.

Gulls are not all *sea* gulls by any means. A vessel headed northward from Seattle will be attended by a cloud of gulls, as vessels are the world over, and in this case the flock will be mostly of the Glaucous-winged Gull, the common and characteristic species of the northwest coast of Washington to the Alaska Peninsula. This is pre-eminently a saltwater bird, and rarely seen far from the coast, though it does occasionally ascend some of the salmon streams for many miles. When a traveler leaves saltwater at Skagway and crosses the White Pass to the lake region beyond, he finds that the streams he travels upon are likewise trailed by flocks of gulls, and it is only the occasional close observer who sees that the personnel of the assemblage has entirely changed. These are mostly the Short-billed Gull, a smaller species than the Glaucous-winged, and abundant throughout the interior of the northwest. The Glaucous-winged soars, circles and sails alongside or behind the vessel it follows. The Short-bill is somewhat different in action in following the lake steamers. The flock trails along pretty steadily for long periods, the bird frequently uttering a rather harsh, three or four syllabled call note. Then, one, two, or three individuals will sail to one side, alight, dip up a drink, and remain there resting. When nearly lost to sight they will arise and overtake the boat again, while in the meantime, perhaps, several others will have dropped out. And this goes on, again and again.

Another breeding gull of the interior is Bonaparte's Gull, a still smaller bird and with (in summer) a conspicuously black head. This species cares nothing for steamers and never follows them. Like the Short-bill in general habits, it breeds on islands in lakes, perhaps three or four pairs in fairly close companionship, though not in colonies, and it bitterly and noisily resents any intrusion during the nesting season. Both the Short-billed and Bonaparte's gulls find their southern breeding limit in the northwest in extreme northern British Columbia. Both migrate commonly along the coast, Bonaparte's often in compact flocks, swift-flying and almost like waders in their direct flight.

The Herring Gull, too, breeds throughout the interior, but this is a cosmopolitan species, and, although not nesting on the coast of the Sitkan district, it is found on saltwater elsewhere.

The Arctic Tern is another cosmopolite. It has attained some renown as the bird that enjoys the most daylight of any, for its circumpolar northern nesting habitat with unending summer daylight is abandoned in the fall for a long flight to the sun-lighted Antarctic region. The Arctic Tern has its southern breeding limit on the lakes of extreme northern British Columbia, where it is abundant; on the coast there is at least one colony in the Sitkan district, at the mouth of the Taku River, not far from Juneau.

Another group of ocean birds of the Alaska coast is found in the petrels, represented by two species, the Fork-tailed Petrel and Beal's Petrel, both nocturnal in habits and unlikely to be found without special search. The group of petrels is cosmopolitan in its distribution, but confined to saltwater; no species is known to occur on even the largest bodies of freshwater.

The beautiful violet-green Pelagic Cormorant is common on the Alaska coast, but it is again a saltwater species purely. The Double-crested Cormorant is common on freshwater over the greater part of the United States and southern Canada, but curiously it does not extend to the apparently suitable lake region of northern British Columbia. A further rather incomprehensible quirk in its distribution is found in the fact that a saltwater representative of this species (the White-crested Cormorant) breeds on the Alaska coast north and west of the region here under consideration, though it is not known to nest in the Sitkan district.

Long-legged Wading Birds and the Coot

There is but one species of heron that is found as far north as Alaska, the Northwest Coast Heron, so called, a local dark-colored variety of the Great Blue Heron, which in one form or another extends over much of North and South America. This bird is practically restricted to the coast, where it is resident, rarely extending inland along the larger rivers, though occasionally seen in summer a hundred miles or so from the sea. The American Bittern has been seen as far north as the Skeena and Stikine rivers on only one or two occasions. The Little Brown Crane is a migrant along the coast and inland, uncommon in both sections. As with the geese, the main line of travel probably lies some distance east of Atlin. The Coot is the surest straggler this far north, reported once or twice from points on the coast, and once or twice from Atlin.

Ducks and Geese

Most of the North American ducks are of wide distribution, many of them, in fact, common to the Old World as the New, and in that part of the Pacific Northwest that is here dealt with there is a very fair representation of the family, though no such great nesting grounds as are known elsewhere in the north. There are but few species that are confined solely to the coast or to the interior, but for the most part there are not a great many freshwater ducks of any kind nesting on its coastal islands. The American Merganser and the Mallard are

Ducks and Geese.

Coast	Interior
American Merganser	American Merganser
Red-breasted Merganser	Red-breasted Merganser
	Hooded Merganser
Mallard	Mallard
	Baldpate
	Green-winged Teal
	Northern Pintail
	Lesser Scaup
	Barrow's Goldeneye
Harlequin Duck	Harlequin Duck
	White-winged Scoter
Canada Goose	

common throughout the Sitkan district, but such birds as Green-winged Teal, Pintail, and Lesser Scaup, abundant in migration and in places throughout the winter, nest for the most part as scattered pairs in the occasional favorable localities.

Freshwater ducks that occur in summer in the interior of northern British Columbia are American Merganser, Red-breasted Merganser (here at its southern summer limit), Hooded Merganser (at its northern limit), Mallard, Baldpate, Green-winged Teal, Pintail, Shoveler (near its northern limit), Lesser Scaup, and Barrow's Golden-eye. Lesser Scaup and Barrow's Golden-eye are *the* common species of the Cassiar district, found on every pond. Canvasback, American Golden-eye, and Bufflehead are known to breed at points north and south of Atlin, but not in that region.

The Harlequin Duck and the White-winged Scoter are noteworthy as maritime species primarily, which seek freshwater in the interior during the nesting season. The Harlequin during most of the year is found along rocky shores, in the most turbulent of coastal waters, surroundings that are deserted in the nesting season for hardly less turbulent mountain streams often at high altitudes and sometimes hundreds of miles from the coast. It probably nests also along streams on the coastal islands. The White-winged Scoter flocks to the inland lakes to nest, often, where the birds are numerous, in semi-colonial manner. The Oldsquaw has been found nesting at Log Cabin, in White Pass, and may do so on Atlin and Tagish lakes. It migrates by the thousands up and down the coast and in lesser numbers inland.

In the Sitkan district there is a species of goose, the White-cheeked Goose (the name is not very distinguishing), a local variety of Canada Goose, that remains in that region throughout the year, sending but a small representation a short distance south of the breeding range in winter, and this bird is rarely found out of sight of saltwater. The meadows at the head of any canal are sure to be occupied by a few pairs, and in some favored localities, such as the famous "flats" at the mouth of the Stikine, many hundreds may be found gathered. In the Cassiar district of the interior there are very few geese nesting, only an occasional pair at long intervals, exact sub-specific status unknown, as there are no breeding birds from that region in collections. It may very likely be the Hutchins' Goose.

Shorebirds

Clouds of waders and of many species pass north and south along the Sitkan coast, but it is not a great breeding ground for such birds. Neither are there more than one or two species that are absolutely restricted to the coast. In the interior, on the other hand, there is quite a variety of waders among the summer visitants, but it is very likely that most of these species nest in limited numbers in the coastal region as well. We lack definite information. The Black Oystercatcher is found nowhere away from saltwater, and, furthermore, it favors the breaking surf and jagged rocks of the outermost islands. The Greater Yellowlegs is found in summer along the Alaska coast to points north of the Sitkan district, but in the interior it appears to be replaced entirely by the closely similar Lesser Yellowlegs. The widespread Spotted Sandpiper nests indifferently over both regions. The Semipalmated Plover, common in summer in the interior of extreme northern British Columbia, is known to breed on the Queen Charlotte Islands and presumably does so also in suitable parts of the Sitkan district, if such there be.

The list of breeding waders of the interior includes the following: Northern Phalarope (apparently at its southern limit near Carcross, Yukon Territory), Wilson's Snipe, Solitary Sandpiper, Bartramia Sandpiper, and Killdeer (rare, and at its extreme northern limit at Atlin). Too little is known of the nesting of these in this general region to make it safe to lay down definite rules covering manner of occurrence, but at any rate in summer they all are far more common inland than on the coast.

Shorebirds.

Coast	Interior
	Red-necked Phalarope
	Wilson's Snipe
Greater Yellowlegs	Lesser Yellowlegs
	Solitary Sandpiper
	Wandering Tattler
	Bartramia Sandpiper
Spotted Sandpiper	Spotted Sandpiper
	Killdeer
Semipalmated Plover (?)	Semipalmated Plover
	Surfbird (?)
Black Oystercatcher	

The wader migrations on the coast and inland bear sufficiently different aspects to warrant their recognition as two distinct lines of travel. Most overlapping of species is to be explained as due to inland birds wandering down the large rivers to the attractive feeding grounds at their mouths. The migrants that follow the Sitkan coast are drawn mostly, I believe, from the Alaska coast to the northwest, including the Aleutian Islands. Those traveling east of the Coast Range may come mostly from the Yukon Valley and the more northern Alaska coast. Among interesting contrasts one of the most striking is afforded by the closely related Western and Semipalmated sandpipers. Along the Sitkan coast the Western Sandpiper migrates in swarms, as it does along the coast to the southward, as far as Lower California. It is practically unknown from the interior, my own experience including one taken at Atlin, another near Hazelton. East of the Coast Range, as at Atlin, the Semipalmated Sandpiper migrates in fair abundance, as it does southward through the interior of British Columbia. On the Sitkan coast it is a mere straggler.

The Pectoral Sandpiper and Baird Sandpiper are common migrants inland, in much lesser numbers along the mainland coast of the Sitkan district. Species that may be explained as stragglers from the Mackenzie Valley are the White-rumped Sandpiper (one specimen, Atlin), Buff-breasted Sandpiper (one specimen, Lake Teslin), and Hudsonian Godwit (one specimen, Atlin).

The Bartramian Sandpiper, known to me only as a south-bound migrant in the upper Skeena region, may also have come from the northeast.

The Long-billed Dowitcher, a common migrant on the coast, is decidedly uncommon inland; and the Red-backed Sandpiper, an abundant coastal migrant, is practically unknown east of the mountains. The Least Sandpiper is the one species that is abundant, and about equally so, in both regions. The Black-bellied Plover is fairly common on the Sitkan coast, rare inland; the American Golden Plover is unknown upon the coast, and rare inland. Occurrences of Sanderling and Hudsonian Curlew are not easily understood. I know of no record of either from the Sitkan district, yet both are common on the coast much farther south. A possible explanation may be direct flight from the Bering Sea region to the southeast, across the ocean to the coast of southern British Columbia on a line that would lie west of the Sitkan district. The Hudsonian Curlew has been taken several times at Lake Atlin and Lake Teslin, the Sanderling once at Atlin.

The Aleutian Sandpiper and the Black Turnstone are characteristic saltwater species that occur in the Sitkan district in clouds, often in flocks together, as transients or as winter visitants. Capture of a single Black Turnstone on Lake Teslin and an Aleutian Sandpiper at Atlin are isolated occurrences that are difficult to explain.

There are two waders to which peculiar interest attaches, the Surf-bird and the Wandering Tattler, both strictly seashore habitants during the non-breeding period, the greater part of the year, and both seek high altitudes, above timberline, to rear their young. This is a striking and unusual seasonal distribution that is a matter of recent discovery. The Surf-bird has been found nesting thus far only on the slopes of Mount McKinley, Alaska, and has been seen in migration at one other place, at Carcross, Yukon Territory. The Wandering Tattler has been found in summer at several points, at Prince William Sound on the Alaska coast (northwest of the region here considered), in the interior of Alaska, and on inland mountains south of the Stikine River.

In reviewing the long list of waterbirds of the northwest, it is apparent that any division of species, of the Sitkan district and of the interior, respectively, is not (with the exceptions of a few) to be based upon, or explained by, any features of topography or climate that are peculiar to these regions. The seabirds (loons, murres, petrels, cormorants, etc.) are species that are wide ranging, many of them cosmopolitan, but upon saltwater mostly or altogether; the freshwater ducks breeding abundantly inland and scarcer on the coast, are again wide-ranging species that merely fail to find any extensive areas on the coast that are suited to their needs. The same appears to be true of many of the waders. Among the gulls, however, there are locally restricted

species, upon the coast and inland, one replacing the other in closely compa-
rable surroundings. Then, the White-cheeked Goose is a good example of a
strongly marked local variety of a widespread species that is closely confined
to its own restricted habitat by intangible but obviously potent barriers. There
is nothing in the character of the interior to inhibit the common presence of
the species, the Canada Goose, but there is something that emphatically does
bar the presence of this particular variety of the Canada Goose, the White-
cheeked Goose.

Passing from waterbirds to the more numerous landbirds, very different conditions are found to prevail in manner of occurrence. Even among the
waterbirds, widespread as most of them are, different species predominate so
markedly in one region or the other as to give a decidedly different aspect to
the aquatic avifauna of the two sections, but of the landbirds there are very
few indeed that are found both on the coast and in the interior, and hardly one
that is equally abundant in both places. It is not too much to say that in the bird
life, particularly the landbirds, as in the plant life, east and west of the dividing
Coast Range (a barrier from thirty to sixty miles wide), the observed differences are as great, and of about the same sort, as those between California and
Minnesota (or even Maine). Taking up the landbirds, group by group, it will
be seen that throughout the entire list there are replacements of closely related
forms in the two sections.

Grouse and Ptarmigan

Of the "Blue Grouse" aggregation (genus *Dendragapus*), inland there is
Fleming's Grouse, a northern variant of the Dusky Grouse of the southern
Rocky Mountains. In the coastal district the Sooty Grouse occurs on the
mainland, this being the northern limits of a habitat that extends southward
over Vancouver Island and into northern California. On the coastal islands is
the Sitka Grouse, a variety of the Sooty, the female of this form being bright
ruddy chestnut in general coloration and one of the most beautiful of North
American grouse. Other varieties of the Sooty Grouse extend southward as
far as southern California. It will be noted that there are here two well-defined
forms (Dusky Grouse and Sooty Grouse), which as several subspecies, extend
north and south, the one from Arizona to the Yukon, the other from southern
California to Glacier Bay, the Coast Range between.

Then there is the Spruce Grouse (genus *Canachites*), which in one form
or another extends across subarctic North America, from Atlantic to Pacific.
A northwestern subspecies, the Alaska Spruce Grouse, ranges southward into
British Columbia to a point about midway between the upper Stikine and
Skeena rivers. A southwestern variety (regarded as a distinct species), Franklin's

Grouse, ranges northward through British Columbia to the same point. The Alaska Spruce Grouse is represented upon the coast, in the northern part of the Sitkan district, by a faintly distinguished subspecies, the Valdez Spruce Grouse, apparently in scanty numbers throughout its range. Franklin's Grouse, of the interior in southern British Columbia, is known to reach the coast in the southern part of the Sitkan district, on Zarembo, Prince of Wales, and Dall islands, and perhaps on others. Island birds are believed to be slightly differently colored from the inland race, but there are very few specimens extant in any museum, too few to settle this point.

The general distribution of *Canachites* is thus very different from that of *Dendragapus*. The latter is confined to western North America, its distribution extending north and south. *Canachites* extends east and west across the continent, and, barely reaching the shores of the Pacific, has there, perhaps, begun to develop newly characterized subspecies in that region, a region that has so potently affected the appearance of nearly all forms of animal life within its bounds.

Blue Grouse and Spruce Grouse are birds primarily of the coniferous forests, but the Ruffed Grouse (*Bonasa*) is found almost entirely in deciduous

Grouse and Ptarmigan.

Coast	Interior
Sooty Grouse (*Dendragapus f. fuliginosus*)	Dusky Grouse (*D. obscurus flemingi*)
"Sitka" Sooty Grouse (*Dendragapus f. sitkensis*)	
"Alaskan" Spruce Grouse (*Canachites canadensis atratus*)	Spruce Grouse (*C. c. osgoodi*)
"Franklin's" Spruce Grouse (*C. c. franklinii*)	"Franklin's" Spruce Grouse
	"Gray" Ruffed Grouse (*Bonasa u. umbelloides*)
Willow Ptarmigan (*Lagopus l. alexandrae*)	Willow Ptarmigan (*L. l. albus*)
Rock Ptarmigan (*Lagopus rupestris dixoni*)	Rock Ptarmigan (*L. r. rupestris*)
	White-tailed Ptarmigan

woods. Its habitat is indicated by the local name in use throughout the northwest, Willow Grouse, its favorite haunts the lowland thickets of willow and poplars. In its several subspecies the Ruffed Grouse extends across North America from ocean to ocean, somewhat farther south than the Spruce Grouse. The northwestern subspecies, the Gray Ruffed Grouse, occupies the interior of British Columbia and Alaska, and another western subspecies, the Oregon Ruffed Grouse (an extremely reddish-colored bird), is found on Vancouver Island and the adjacent mainland. It is noteworthy that neither of these forms reaches the coast of southeastern Alaska, though at first thought it would seem reasonable to expect to find the Oregon Ruffed Grouse ranging northward into that region. The explanation probably lies in the absence of any considerable amount of deciduous woods in the Alaska coastal belt, the nearest substitute, the relatively restricted thickets of scrubby willow and alder, probably not filling the bird's needs.

The remaining grouse of the northwest, the several species of Ptarmigan, are birds of many unusual aspects of plumage, habits, and life history, so much so that it is difficult to avoid here elaborating upon various of these features, irrelevant to the distribution of the several forms. Mention should be made, though, of one striking character of the ptarmigans, namely their seasonal changes of plumage from brown or gray in summer, to white in winter.

There are three species of ptarmigan in North America, Willow Ptarmigan, Rock Ptarmigan, and White-tailed Ptarmigan, each of which has one or more local subspecies in the northwest. The Willow Ptarmigan is circumpolar in distribution, ranging across northern Europe, Asia, and North America. The Rock Ptarmigan, very closely similar to the Old World *Lagopus mutus*, is the northernmost of the three, extending far toward the pole. The White-tailed Ptarmigan is purely American, being confined to the mountains of the west. The Willow Ptarmigan is represented by different subspecies in the interior and on the coast, and the Rock Ptarmigan, too, has different subspecies inland and on the coast. The White-tailed Ptarmigan is only in the interior. The Willow Ptarmigan (*Lagopus albus*) inland occupies during the summer the lower mountain slopes and valleys immediately above timberline, down to about the 3,000-foot contour. It ranges southward at least as far as the upper Skeena River and the Yellowhead Pass. The coastal variety (*Lagopus lagopus alexandrae*) is found at sea level in the northern part of the Sitkan district, but farther south (on Prince of Wales and adjacent islands) it remains almost entirely on the mountaintop, largely, perhaps, on account of the absence of suitable open country at lower levels. Its southern limit on the coast is at about the same latitude as the inland subspecies.

The Rock Ptarmigan (*Lagopus rupestris rupestris* inland, *L. r. dixoni* on the coast) is a bird of high altitudes in both regions, on bleak, wind-swept peaks and ridges on the Alexander Archipelago, on high elevations, but not rugged character, in the interior. Inland it extends southward, not commonly, at least as far as certain mountaintops near Hazelton. On the coast, *dixoni* is more restrictively northern, having been found only on Baranof and Chichagof islands and on the adjacent mainland.

The White-tailed Ptarmigan (*Lagopus leucurus*) is of interest on several counts. It is purely North American, whereas the other two species range clear round the globe. It also displays two apparently anomalous characters, as follows: It is the most highly specialized of the ptarmigans in its entirely white winter plumage, for while *lagopus* and *rupestris* have coal-black tail feathers, summer and winter, in *leucurus* the tail feathers are white at all times. Thus, in winter plumage it is pure white throughout, exemplifying the extreme development of the white seasonal plumage. At the same time, instead of being one of the most northern of birds, a feature that might have been anticipated as an accompaniment of such extremely specialized winter plumage, it is by far the most southern of the New World ptarmigan, ranging along the Coast Range and Cascades to Vancouver Island and Mount Rainier, along the Rocky Mountains to the northern boundary of New Mexico. In the north, as in the south, the White-tailed Ptarmigan frequents high altitudes and the most bleak and precipitous surroundings, far more so than the Rock Ptarmigan. In fact, it goes some places by the name of the "Rock Ptarmigan," from the nature of its habitat, while the true Rock Ptarmigan is either confused with the larger Willow Ptarmigan or else known as "Croaker," from its call notes. Curiously, the White-tailed Ptarmigan does not occur on the coast of southeastern Alaska or of British Columbia north of Vancouver Island. Perhaps it is better to modify that statement to the effect that the species has never yet been found in that section, but the region is rugged and explorers have been few. At any rate, the other two species of ptarmigan have been found there, but not the White-tailed.

In summarizing the manner of distribution of the grouse of the north it will be noted that both in the interior and on the coast there are some species at low levels, others at higher elevations. It will be noted that there are closely related (but distinct) subspecies occupying respectively the lowlands of each of the two regions, other closely related (but distinct) subspecies at the higher altitudes. There is, however, still another feature to the distribution of one of the groups concerned: the Blue Grouse (*Dendragapus*) on the coast is a bird of the lower levels, occurring at sea level. In the interior it is found breeding only at high altitudes, about at the timberline level, and it only sporadically descends into the lowlands at the end of summer.

Hawks, Eagles, and Falcons

Hawks generally nest as widely scattered pairs, and most species are extremely secretive and quiet during the breeding season, so that it is sometimes hazardous to venture upon generalizations regarding their distribution. A comparative list, as given here, is bound to invite attack, more or less justified. It is nevertheless useful for illustrating certain points I wish to make, and it may stand therefore with all its imperfections. Many of the species listed require modifying or explanatory comments in support of their position in one or the other column, but there are several that are excellent examples of the sort of differences commonly prevailing between the two regions.

The Marsh Hawk is primarily of the interior, and I do not know that it has ever been found nesting in the Sitkan district, where it is a common migrant. The Sharp-shinned Hawk, wide ranging across the continent, is apparently in both regions. The status of the Goshawk is debatable. For years

Hawks, Eagles, and Falcons.

Coast	Interior
	Marsh Hawk
Sharp-shinned Hawk	Sharp-shinned Hawk
Goshawk (*Astur atricapillus striatulus*)	Goshawk (*Astur a. atricapillus*)
"Western" Red-tailed Hawk	"Harlan's" Red-tailed Hawk
	Swainson's Hawk
"Northern" Bald Eagle	
	Golden Eagle
	Gyrfalcon
Peregrine Falcon	Peregrine Falcon
Pigeon Hawk	Pigeon Hawk
	Sparrow Hawk
Osprey	Osprey

there was general belief in the existence of a dark-colored coastal subspecies peculiar to the northwest, a belief that is at present suffering some criticism. At any rate, the Goshawk is far more abundant in the interior than on the coast, though in any given region its numbers fluctuate widely from year to year.

The Red-tailed Hawks divide in a much more satisfactory manner. In the Sitkan district is the Western Red-tailed Hawk, of wide distribution farther south, from the Rocky Mountains to the Pacific. From central British Columbia northward the habitat of the Western Red-tail narrows to an attenuated strip along the coast, throughout the Sitkan district, where, however, this Red-tail is a decidedly uncommon bird. East of the Coast Range it is abruptly replaced by the strikingly different Harlan's Hawk, almost coal black and with the tail marked with whitish and practically destitute of red. This is a bird with an interesting history. Discovered by Audubon in Louisiana, it was until recently believed to be restricted to the Gulf Coast, in reality its winter home, while many ornithologists were outspoken in their disbelief in its existence at all as a subspecies. It is, however, a far northern variety of the Red-tail, summering in adjacent parts of Alaska, Yukon Territory, and British Columbia, and migrating southeastward in the fall, east of the Rocky Mountains and down the Mississippi Valley.

Swainson's Hawk, too, "stays put" geographically, in a very satisfying manner, absolutely eschewing the coast and restricting itself to the interior, and to the open country thereof, which means usually above timberline.

The Northern Bald Eagle probably extends across the continent, but it is so characteristically a bird of the coast, and is so rare in the northern interior of British Columbia, that it seems proper to list it among the coastal birds. There are probably a pair or two of Bald Eagles to be found about any large inland lake, where fish are to be had, but on the Alaska coast, before the recent depletion of the eagles from the bounty placed on them, they existed in numbers that were well-nigh incredible for such a large predatory bird. The Golden Eagle, on the other hand, cosmopolitan in general distribution, is here closely confined to the eastward of the Coast Range. It is, I believe, absolutely unknown upon the sea coast, though not uncommon inland.

The Duck Hawk, or Peregrine Falcon, another cosmopolite that varies slightly in appearance in different parts of the world, can hardly be safely restricted to one column or the other. It probably nests in both regions, though not commonly in either. Its discovery actually nesting in the Sitkan district is of especial interest in that another strongly marked form of the Peregrine, Peale's Falcon (*Falco peregrinus pealei*), apparently replaces it entirely in the Queen Charlotte Islands to the southward, and on the Alaska islands to the northwest, from Kodiak to the shores of Asia. It may be said that the apparently discontinuous breeding range of the Peale's Falcon does not look so singular when it is plotted on a globe, for the Aleutian Islands are then seen to lie almost directly west of the Queen Charlottes and scarcely any farther north, and the Sitkan district is not the intervening territory that it would seem to be. So, Peale's Falcon, despite the positive statements in many books of its nesting along the whole Alaska coast, apparently leaves the Sitkan district to the occupancy of its cousin, the Duck Hawk.

The Gyrfalcon is a boreal species that in summer extends southward barely into the northern interior of British Columbia, a rare inhabitant there of the open country above timberline.

The Pigeon Hawk is so secretive in its nesting that in several years of field work in the northwest I have never so much as seen one of the birds during the nesting season. Toward the end of the summer Pigeon Hawks usually appear in some numbers, both on the coast and inland, migrating southward. This is a species that extends across the continent in its general distribution, breeding assuredly in the northern interior. Whether or not it actually breeds on the coast of southeastern Alaska (the Sitkan district), where it is a common migrant, I do not know. There is an extremely dark-colored form of the Pigeon Hawk, so distinctive in appearance that it has been granted a separate name, the Black Merlin (*Falco columbarius suckleyi*). It has been assumed that this is a northwest coast variety, of dark color as are so many of the birds of that region, and entirely replacing the true Pigeon Hawk in the humid coastal belt. The nesting grounds of the Black Merlin have never been found, however, nor, I believe has the bird itself ever been taken on the coast north of Vancouver Island, while south-bound migrants have been shot in the interior as far north as Atlin. It is thus at least an open question as to whether the so-called Black Merlin may not be a color phase of the Pigeon Hawk (just as there are dark-colored and pale-colored Swainson's Hawks and Rough-legged Hawks), but a color phase perhaps of limited geographical distribution, possibly in the northwestern interior. It would not be the first time that a color phase was dignified by a separate name.

The Sparrow Hawk shuns the coast north of Vancouver Island, but extends into Alaska in the interior. It is a bird that prefers open country or a region of scattered trees, and it may be that its appearance on the coast is barred by the dense forests of the region rather than directly by any climatic feature. I have seen one or two stray individuals at points on the mainland coast, as at the mouth of the Stikine River, during migration at the end of summer, but the Sparrow Hawk is only a rare straggler at that section.

The Osprey is widespread over North America, and it occurs both inland and on the coast, though nowhere in abundance. It is apt to be found about any large body of water where there are good-size fish to be caught.

Owls

If we must beg off from supplying exact information regarding the distribution of certain of the hawks, it will be understood that the nocturnal owls offer still greater difficulty in plotting the precise limits of their distribution. Here again, however, we can to some extent discern the workings of distributional laws such as affect most of the animal and plant life of the regions.

The Short-eared Owl nests far to the southward and far to the northward of this part of the country, but it shuns the heavily wooded coast. The Long-eared Owl is a southern bird, extending northward regularly as far as the interior of central British Columbia. The collecting of a single bird by myself

Owls.

Coast	Interior
	Long-eared Owl (*Asio wilsonianus*)
	Short-eared Owl (*A. flammeus*)
	Great Gray Owl (*Scotiaptex n. nebulosa*)
	Boreal "Richardson's" Owl (*Cryptoglaux funerea richardsonia*)
Northern Saw-whet Owl (*Cryptoglaux acadica*)	Northern Saw-whet Owl (*C. acadica*)
"Dusky" Great Horned Owl (*Bubo virginianus saturatus*)	
	"Arctic" Great Horned Owl (*B. v. subarcticus*)
	"Northwestern" Great Horned Owl (*B. v. lagophorus*)
	Northern Hawk Owl (*Surnia ulula caparoch*)
Northern Pygmy-Owl (*Glaucidium gnoma grinnelli*)	

on the lower Taku River, near the Alaska–British Columbia boundary, in September 1909, is an isolated occurrence the meaning of which is not yet clear, as the species has never been found in the adjacent interior nor elsewhere on the coast.

The Great Gray Owl, Richardson's Owl, and the Hawk Owl are all of them birds of dense woods, but still interior species solely; they have not yet been found in the coastal forest.

The Acadian Owl presents some peculiar points. It is subarctic (that is, Canadian and Hudsonian life zones) in its distribution, extending across the continent from ocean to ocean, even into the Sitkan district. On the Queen Charlotte Islands there is what appears to be a perfectly distinct (though closely related) species, Fleming's Owl (*Cryptoglaux flemingi*), confined to those islands; the Acadian Owl apparently visits the islands during the winter months.

The Great Horned Owl, responsive as it is to varied local conditions throughout its extensive habitat, occurs as different subspecies on the coast and inland. The dark-colored coastal form, the Dusky Horned Owl, is found throughout the Sitkan district. Inland are two paler-colored forms. The Arctic Horned Owl, from the Cassiar district northward, is a very whitish bird. In the Omineca district there is a more brownish-colored form, the "Northwestern Horned Owl" (a manufactured "book name," if ever there were one). In the interior, one or the other of these two subspecies are periodically and locally of great abundance, depending mostly upon the condition of the rabbit population. The Dusky Horned Owl is the only species of owl that is of even fair abundance on the coast.

We know little about the Pygmy Owl in this region. It does occur as far north as the Stikine River, both inland and at the mouth of that stream, but it is a rare bird at both places. It has been recorded from Atlin and on the coast from Wrangell. The Wrangell occurrence may have been the result of migration from the interior down the Stikine Valley. I know of no coastal records between that point and Vancouver Island.

Nighthawk, Swifts, and Hummingbird

The same Nighthawk that raises its broods upon the graveled roofs of smoky cities on the Atlantic coast does so also in the open woodland of northern British Columbia. There it does not pass beyond the Coast Range, and it is unknown in the Sitkan district, but it is common over the southern mainland of British Columbia and even over Vancouver Island.

Black Swift and Vaux Swift both extend northward to about the same latitude in the Cassiar district and in the Sitkan district. Both are common in the Skeena Valley, and uncommon in the upper Stikine Valley, apparently their

Nighthawk, Swifts, and Hummingbird.

Coast	Interior
	Eastern Nighthawk
Black Swift	Black Swift
Vaux Swift	Vaux Swift
Rufous Hummingbird	Rufous Hummingbird

northern limit. The Black Swift is abundant on the Alaska coast at least as far as the mouth of the Stikine, and Vaux Swift, much less numerous, extends about as far north. Neither is found on any but the innermost islands. The Rufous Hummingbird is common in the upper Skeena Valley, and decidedly rare in the Atlin region, where one may see, perhaps, two, three, or four birds during a summer. It ranges north along the Alaska coast through the length of the Sitkan district, so there is little difference in manner of occurrence on the two sides of the Coast Range.

Belted Kingfisher

The Western Belted Kingfisher (*Ceryle alcyon caurina*) is a very slightly differentiated western subspecies of a wide-ranging North American bird, notable in that it occurs in equal numbers and in unchanged appearance, inland and on the coast.

Woodpeckers

The family of woodpeckers is well represented in the northern woods, and there are various species that are characteristic of and fairly common in the several regions here considered. The widespread Hairy Woodpecker occurs as three subspecies. The large Northern Hairy Woodpecker has its southern limit in the Atlin region, where it is rather rare. The Rocky Mountain Hairy Woodpecker reaches its northern limit in the valley of the Stikine, and it is fairly abundant from that region southward. The Sitka Hairy Woodpecker is confined to the Sitkan district. All three of these are white-breasted birds, as distinct from the sooty-breasted subspecies to the southward. It is a noteworthy fact that the Sitkan Woodpecker is an offshoot of the interior Rocky Mountain Hairy Woodpecker, and not closely related to the dark-colored Harris Woodpecker of the coast of southern British Columbia, with which it was once thought to be identical. The avifauna of the Sitkan district is mostly derived from the coast to the southward, but there are some elements, and

Woodpeckers.

Coast	Interior
"Sitka" Hairy Woodpecker	"Northern" Hairy Woodpecker
	"Rocky Mountain" Hairy Woodpecker
	"Nelson's" Downy Woodpecker
"Rocky Mountain" Downy Woodpecker	"Rocky Mountain" Downy Woodpecker
	Arctic Three-toed Woodpecker
"Alaska" Three-toed Woodpecker	"Alaska" Three-toed Woodpecker
	Yellow-bellied Sapsucker
Red-breasted Sapsucker	Red-breasted Sapsucker
	"Western" Pileated Woodpecker
"Red-shafted" Northern Flicker	"Yellow-shafted" Northern Flicker

this woodpecker is one of them, that unquestionably had their origin in the adjacent hinterland to the eastward.

Conditions in the Hairy Woodpecker group are practically duplicated in the smaller Downy Woodpecker. The northern Nelson's Downy Woodpecker occurs, but is not common, in the Atlin region, its southern limit. I saw no Downy Woodpeckers in the Stikine Valley, though the species must occur there, but found the Rocky Mountain Downy Woodpecker to be fairly numerous in the Skeena Valley, and it is a common bird from there southward. This subspecies, too, reaches the coast of the Sitkan district, and even farther westward to Prince William Sound, but not in any numbers and perhaps only as a migrant or wanderer. Again, paralleling the non-occurrence of the Harris Woodpecker, the dark-colored Gairdner's Woodpecker, of the coast of southern British Columbia, does not extend northward into the Sitkan district. So that, while the Hairy Woodpecker of the interior has established a "colony" on the coast, which has been there long enough to develop distinguishable characters of its own, the Downy Woodpecker appears to be still at an earlier stage, with only a few adventurous explorers trying out the more nearly adjacent portion of a region that (perhaps as yet but poorly adapted to the species' needs) may in time be thoroughly colonized.

In the Three-toed Woodpeckers, the Arctic or Black-backed (despite its name the most southern species of the group), widespread across North America and far south in the Rocky Mountains and Sierra Nevada, occurs throughout the interior but has not yet been found upon the coast. The Alaskan Three-toed Woodpecker (a subspecies of the American or Ladder-backed) is fairly common inland, and, like the Downy Woodpecker of the same section, has established a precarious foothold in the Sitkan district, where it has been seen once or twice.

The Red-breasted Sapsucker is a Pacific coast species, in California ranging from the west slope of the Sierra Nevada westward, and extending northward in a narrowing coastal strip to its northern limit in the Sitkan district. A coastal bird for the most part, it ranges inland up the Stikine River barely to the east side of the mountains, and no farther. In the broad Skeena Valley conditions are different, and the Red-breasted Sapsucker is common there eastward to Hazelton and beyond, one of the very few coastal birds that are establishing themselves far from their normal habitat. The Yellow-bellied Sapsucker of eastern North America has been found at Telegraph Creek, apparently the extreme northwestern point of its range to which it has attained.

The flickers in their distribution present several features worth dwelling upon. The range of the Yellow-shafted Flicker of eastern North America (from the Rocky Mountains eastward in the United States), here in the northwest reaches westward to the east base of the Coast Range, only a few miles from the coast. The range of the Red-shafted Flicker of western North America (from the Rocky Mountains westward in the United States) is here correspondingly narrowed to the attenuated coastal strip comprising the Sitkan district. As these two species are known to hybridize rather freely where conditions permit, the fact that throughout the Cassiar district and over most of the Sitkan district their specific characters are retained in all purity is a strong testimony to the efficacy of the faunal barrier between.

In the Skeena Valley of the Omineca district conditions are different. As shown from other evidence, coastal conditions, much modified, extend far inland up this broad valley, and the flickers give witness to this also. The Yellow-shafted Flicker is the species of the Hazelton region, but practically the whole flicker population of that section, 200 miles inland, shows traces of Red-shafted ancestry. There are almost none of pure blood. There is a notable contrast between the appearance of the mongrel breeding stock (predominantly Yellow-shafted in character, however), and the pure-blooded birds that appear from the north in the fall migration. Then, on Revillagigedo Island, Alaska (immediately north of the mouth of the Skeena River), in the habitat of the Red-shafted Flicker, I once took a pair of breeding birds, the male of which was a hybrid, but predominantly of Yellow-shafted characters. There is, however, apparently far more invasion of the Red-shafted species inland than of the Yellow-shafted species toward the coast.

The big Pileated Woodpecker, transcontinental in distribution and ranging over Vancouver Island in southern British Columbia, here in the north does not extend west of the Coast Range. It occurs northward regularly to the Skeena River and occasionally even to the vicinity of Telegraph Creek. It is of interest here merely as being at the extreme northwestern corner of its entire habitat, checked apparently by the barrier of the Coast Range.

Flycatchers

The Western Flycatcher is the only species of this assemblage that is characteristic of the Sitkan district, or even fairly common therein. This is a wide-ranging species that occurs in unchanged appearance from Mexico to Alaska, but in the north, from southern British Columbia northward, it is rigidly confined to the narrow coastal strip, and never seen inland. The characteristic and penetrating call note of the Western Flycatcher is to be heard in the woods of even the westernmost islands of the Alexander Archipelago, where it may be almost the only bird sound to break the stillness.

Three other flycatchers, the Alder Flycatcher, Olive-sided Flycatcher, and Western Wood Pewee, have all been found (all decidedly rare) at several points on the mainland coast of Alaska, but, although they may breed regularly at such places, these birds merely represent extreme outposts of the species

Flycatchers.

Coast	Interior
	Eastern Kingbird
	Say's Phoebe
Olive-sided Flycatcher	Olive-sided Flycatcher
Western Wood Pewee	Western Wood Pewee
	Yellow-bellied Flycatcher
Western Flycatcher	
Alder Flycatcher	Alder Flycatcher
	Hammond's Flycatcher
	Wright's Flycatcher

that have been extended from the interior habitat through favorable avenues into the confines of the west belt. Their presence there, however, may hold a promise of a future wider range.

The eastern prototype of the Western Flycatcher, the Yellow-bellied Flycatcher, has been taken once near Hazelton and once near Atlin, but it is apparently only a straggler so far to the westward. Three other species of *Empidonax* fill the niches inland that are occupied by the Western Flycatcher on the coast. The Alder Flycatcher occupies the willow thickets of the lowlands (in the absence of alders in much of the interior, perhaps), and the Hammond (commonly) and Wright's (rarely), the heavier woods. The Western Wood Pewee, too, is common in the more open lowland woods, and the Olive-sided Flycatcher, though not abundant, is of general distribution throughout the interior. The Say's Phoebe, nowhere common, is almost sure to be found about human habitations, occupied or abandoned, to points far north of the regions here considered. The Eastern Kingbird reaches its northwestern confines about Hazelton, where it is of extreme rarity.

Horned Lark

In more southern regions Horned Larks are found exclusively in the lowlands, characteristic birds of prairie, field, and pasture. In the forested subarctic regions the only suitable open country for the Horned Larks lies above timberline, and this means as a rule the ridges and plateaus of the higher mountains. The Pallid Horned Lark (*Otocoris alpestris arcticola*) is found throughout the interior in Alpine surroundings, an associate or a neighbor of the Rock Ptarmigan and the mountain sheep. It is not known to occur anywhere in the Sitkan district.

Swallows

Swallows are rather rare in the coastal region. Only two species occur there in any numbers, the Barn Swallow and the Tree Swallow, and even they are not nearly so abundant there as in the interior. Also, they are mostly along the mainland coast and on the inner islands, of relatively rare occurrence farther west. The swallows nearly all thrive about human habitations, and on the coast the Barn Swallow, in the interior Cliff, Barn, Tree, and Violet-green swallows, all seek the small towns and even the scattered single cabins, where they nest in whatever nooks and crannies are available, or in bird houses, eagerly occupied wherever they are set up.

In Telegraph Creek I had pointed out to me a Barn Swallow's nest, on a ledge by a window, that had been added to, year after year, until it resembled a story-book picture of a Flamingo's nest. At the Portage near Atlin a pair of Barn

Swallows.

Coast	Interior
	Cliff Swallow
Barn Swallow	Barn Swallow
Tree Swallow	Tree Swallow
	Violet-green Swallow
	Bank Swallow
	Rough-winged Swallow

Swallows, summer after summer for many years, occupied a nest in a box attached to the roof of a railroad car, a car that traveled back and forth across the Portage at frequent intervals. The Cliff Swallow has never been taken on the coast. The Violet-green Swallow occurs in small numbers on the Alaska coast, at least in late summer; it is not known to breed there. The Rough-winged Swallow finds its northern limit in the vicinity of Hazelton, in the upper Skeena Valley, while the Bank Swallow goes very far north.

Magpie, Jays, and Crows

Distribution and migrations of the Magpie in the region we are considering are not well understood, but although it reaches the coast far to the northwest of this section, here, and from here southward, it is emphatically a species of the interior. The Magpie breeds at Carcross, some seventy-five miles northwest of Atlin, but is not known to do so near Atlin, Telegraph Creek, or Hazelton, though I believe that at all these points (at Atlin certainly) it turns up pretty regularly in the late summer. At this time, too, it wanders down the Taku and Stikine rivers to the coast.

The crested jays of the genus *Cyanocitta* are American in origin and must have invaded Alaska and British Columbia from the south. In western North America there are two rather widely divergent strains, on the one hand, the several subspecies that occupy the Sierra Nevada and Coast Range, which have no white markings above the eye; on the other, the Rocky Mountain races, which have such markings. The Rocky Mountain subspecies, *annectens* and *diademata*, are northern links of a connected chain extending south through Mexico into Central America. The coastal subspecies form an almost isolated group, widely separated from the Rocky Mountain strain at the south, but with some degree of junction at the north. In British Columbia there is geographic approach coupled with evident intergradation. Existing conditions suggest that the Pacific coast subspecies are the result of a "back-wash" in distribution, a southward advance in the Coast Range and Sierra Nevada, accomplished from some point or points on the northern coast. The ancestral *Cyanocitta* may be supposed to have passed upward from Mexico to its farthest north in the interior, to have eventually reached the coast, and then to have found its way southward in congenial surroundings that were otherwise unattainable. The Alaska coast, now occupied as far as Cook's Inlet, the northernmost point reached by the genus, was probably colonized last in the history of the movement.

Steller's Jay is a characteristic and conspicuous inhabitant of the coastal region from Vancouver Island to Cook's Inlet. In the southern interior of British Columbia is another subspecies, the Black-headed Jay, which reaches its northern limit in the Skeena Valley. The jay population of the Hazelton region, 200 miles from the coast, shows a strong admixture of Steller's Jay characteristics; fairly typical examples of the Black-headed Jay have been taken as far as the mouth of the Skeena River, and even beyond, on the nearby Porcher Island, in the coastal habitat of the Steller's Jay. So, at the mouth of the Skeena there is a break in the north-and-south distribution of the bird we call *Cyanocitta stelleri*, due to the interposition of the *annectens* strain. Immediately to the southward, on the Queen Charlotte Islands, is the extremely dark-colored subspecies *carlottae*. Farther south, on Vancouver Island, and farther north, on the Alaska coast, the birds are less dark, and these two separated but superficially similar strains are classified under the same name. It is a situation that cannot be adequately met by our system of nomenclature.

In the intergradient population between the interior and the coast, the jays of this species present a different situation from that seen in some

Magpie, Jays, and Crows.

Coast	Interior
	Black-billed Magpie
Steller's Jay (*Cyanocitta stelleri stelleri*)	"Black-headed" Steller's Jay (*C. s. annectens*)
	Gray Jay
Common Raven	Common Raven
North-west Crow	
	American Crow

other groups, such as the two subspecies of the thrush *Hylocichla ustulata*. In that case the two subspecies *ustulata* and *swainsoni* meet as distinct species, without intergrading, suggesting that they have approached and met after long separation. The condition in the Steller's Jay variants, with intergradation in intermediate territory, suggests uninterrupted communication through the ages; and it suggests to my mind the northward, then southward, movement of the population that is outlined above.

The Canada Jay is a Boreal species that is transcontinental in its distribution, save that it, again, is stopped within sight of the Pacific coast by the barrier of the Coast Range. It is unknown to the Sitkan district. South as far as the Stikine River the Canada Jay is a dweller of the lowlands as well as of the mountainsides. Farther south, toward Hazelton, it apparently does not breed in the valleys, but only in the Hudsonian zone of the higher slopes, descending therefrom at the end of summer.

The northern Raven is much like the Bald Eagle in its associational preferences, being abundant everywhere along the coast and decidedly rare in the interior. It is, of course, transcontinental (circumpolar, in fact) in its general distribution, but there is the decidedly noticeable local contrast here, of ubiquitous presence in the Sitkan littoral, and marked scarcity east of the Coast Range.

The habitat of the Northwest-Crow is implied in the one term, "beach comber"; it is almost never seen out of sight of saltwater. Along the British Columbian and Alaska coast every bay and inlet has its quota of crows searching the mud flats and beaches, north throughout the Sitkan district. The Western Crow (a slightly distinguished western variety of the common American Crow) occurs in the interior valleys of British Columbia north to the vicinity of Hazelton, but nowhere in this region is it known to reach the coast.

Chickadees and Nuthatch

The vast difference between coast and interior is emphasized in the chickadees perhaps more than in any other group of birds. This is evident in the paucity of species on the coast as compared with the interior, and in the nature of the differences between the most nearly related forms in the two regions. The Chestnut-backed Chickadee (*the* coastal species) is obviously an offshoot from the same parental stock as the Hudsonian Chickadee of the interior. Whether or not the Chestnut-backed species has been established in its present home for any very long period (and, as has been said, it does not seem as though the Sitkan district could have been habitable since extremely remote time), its strongly marked characters point to absolute seclusion upon the coast since its establishment there. The Hudsonian Chickadee has developed several rather weakly indicated subspecies in various parts of its wide habitat across subarctic North America, but here on the western coast in rigid isolation, there is a strongly marked form, a "species," as we insist, that is undoubtedly of the same ancestry and probably of no greater age. The Chestnut-backed Chickadee, it may also be pointed out, is one of the few species of the Sitkan avifauna that is of recent Boreal derivation. Most of the birds, the smaller ones especially, are from the south. The Chestnut-backed Chickadee occupies the whole of the Sitkan district, mainland and islands. It ranges up the Stikine only a few miles; up the broader Skeena Valley it goes at least as far as Hazelton, but not commonly.

The Chestnut-backed Chickadee is a rather unconforming element in the Sitkan avifauna, more so than appears upon first consideration. It and the Northwest Crow are the only well-defined *species* that may be assumed to have had their origin on the northwest coast. The rest of the avifauna consists of more or less strongly marked local subspecies of widespread species. In its sharp specific characteristics the chickadee appears to be of greater age than any of its associates except the crow.

The Chestnut-backed Chickadee is of Boreal origin, and whereas most of the distinctive Sitkan birds are northern variants of southern species, the few of the northern region, other than this one, are not strongly modified from the parent stock. Close resemblance of Chestnut-backed Chickadee and Hudsonian Chickadee is obvious, and close relationship is inferred, but immediate derivation of the Chestnut-backed from the Hudsonian stock does not seem to have been possible—unless specific differentiation in the Chestnut-backed Chickadee advanced at a far more rapid rate than in other birds. The post-glacial reoccupation of the Sitkan district seems too recent an event to permit it. It would be a remarkable circumstance if this Chickadee became differentiated and specifically isolated in the Sitkan district in the presumably brief period since that region has been habitable, a period that has had so little effect, relatively, upon the appearance of the rest of the fauna.

As to the few species that are known to have reached the coast from the east or northeast, some are very slightly changed from the parent stock, some not changed at all. The Chestnut-backed Chickadee can hardly have been of southern derivation in the first place, yet, judging from the slight differentiation shown by the rest of the scanty Boreal fauna of the Sitkan district, it could hardly have arrived with those species either. The explanation of this curious bird may lie in a more remote past than concerns its present associates, entailing a pre-glacial retreat southward, perhaps, with a long period of isolation, followed by post-glacial reoccupation of the northern coast. It is a striking fact that the two forms, the Chestnut-backed and the Hudsonian, exist in such

Chickadees and Nuthatch.

Coast	Interior
Chestnut-backed Chickadee	"Hudsonian" Boreal Chickadee
	"Long-tailed" Black-capped Chickadee
	Mountain Chickadee
	Canada Nuthatch

close proximity, coastwise and inland, with so little invasion of each other's territory and with no interbreeding.

The Hudsonian Chickadee is a common lowland species as far south as the Atlin region, mostly in the white spruce forests. Farther south, as about Hazelton, the Hudsonian zone, to which this species belongs, climbs to higher altitudes, and the chickadee breeds not lower than the 4,000-foot level. It has nowhere penetrated to the coast. The Long-tailed Chickadee is a northwestern variant of the species (*Penthestes atricapillus*) that in one form or another covers much of northern North America—the common Black-capped Chickadee, *the* chickadee of the East. Throughout the interior of northern British Columbia it is a common and characteristic bird of the quaking aspen woods, shunning the denser spruce forests and remaining in the valleys. It is of wide distribution farther northward, to the limit of trees in fact, but extension to the west is halted abruptly by the Coast Range. It has nowhere reached the coast of southeastern Alaska.

The Mountain Chickadee (occurring over much of North America west from the Rocky Mountains) is represented in British Columbia by a slightly differentiated variant. It is common in the southern part of the province, increasingly rare northward, but recorded from the upper Skeena Valley, the upper Stikine, and at Atlin. This again is an inland species, ranging from northern British Columbia to southern California, but always back from the coast.

It is not possible to define exactly the status of the Canada Nuthatch. It occurs across the continent in the northern woods, but in the regions here discussed I have never found it in summer in sufficient numbers anywhere to satisfy me that it was a characteristic component of the avifauna of that section. I believe that it probably is better placed with the avifauna of the interior (where I have found occupied nests). In the Sitkan district one or two of the birds have been seen when they might have been breeding, but no nests have been found and the species is assuredly rare there in the summer. It migrates rather commonly both on the coast and inland.

Pipit, Dipper, Catbird, Wrens, and Creeper

The Pipit is a common migrant coastwise as well as inland, but so far as known it nests only in the interior. It occupies an Alpine-Arctic habitat over much of western North America, including most of the above-timberline summits and plateaus of the Omineca and Cassiar districts east from the inner side of the Coast Range. There it may be found in company with Rock Ptarmigan, Horned Lark, Golden-crowned Sparrow, and sometime with Hepburn's Rosy Finch.

The Dipper occurs indifferently throughout the Sitkan district and inland, though seldom in any numbers at any one place. Presumably the nature of its habitat along the rushing streams is such as to permit entry into adjacent new territory just as fast as it becomes available. The species is tolerant of a wide range in temperature and humidity, and with every valley and canyon affording habitable surroundings there has evidently been nothing to hinder the spread of the Dipper over this whole northwestern country. It is, moreover, resident the year through, despite the extremely low winter temperatures that in part of this section seal up most of the remaining streams on which its living depends.

The Catbird seems decidedly out of place amid its subarctic associates, but must be included here on the basis of one bird, unmistakably breeding, that was taken near Hazelton, on the upper Skeena River. This is the northwestern limit of the species, common in southern and central British Columbia, but not I think to be regarded as an isolated occurrence, unlikely to be repeated. It seems to me that the Catbird may well be one of the birds that has been following up the spread of cultivation and habitations toward the northwest, eagerly occupying clearings and second growth in the forest, and prepared to

Pipit, Dipper, Catbird, Wrens, and Creeper.

Coast	Interior
	American Pipit
American Dipper	American Dipper
	Gray Catbird
	Western House Wren
Western Winter Wren	
"Tawny" Brown Creeper	

hold its ground in the altered surroundings. The Western House Wren comes into our list on exactly the same basis as the Catbird (one specimen collected near Hazelton), and is probably to be regarded in exactly the same light, again a species to which dense forest is repellant, common a little father south and pushing northward as partly cultivated clearings are available.

Of very different stripe is the Western Winter Wren, habitant of densest forest and dripping surroundings. Primarily of the coast region, Winter Wrens of various subspecies extend from the Sitkan district northward and westward the length of the Aleutian Islands to a close connection with Asiatic relatives. Throughout the Sitkan district the Western Winter Wren is abundant and in all sorts of surroundings. On the eastern slope of the Coast Range it is rare and only to be found at high altitudes, as I have seen it on a mountain near Hazelton (4,000 feet), and above the central Stikine (3,000 feet). I was unable to find the species in the Atlin region, though it may occur on the western side of Tagish Lake.

The Brown Creeper in its several subspecies occurs over most of North America; there are some curious anomalies in its manner of distribution in the northwest. The Tawny Creeper is the subspecies of the humid coast region, a well-marked variety of fairly common occurrence in the Sitkan district and southward to California. It is closely restricted to the region west of the Coast Range, and in the Sitkan district it is found over the mainland and islands, though as with all other small birds in lesser numbers on the westernmost islands. East of the Coast Range, in the Omineca, Cassiar, and Atlin regions is a vast extent of country that is apparently well suited to the Brown Creeper's needs and in which no Brown Creeper is found. I believe that a very few migrants do pass through (I have seen one or two, in fact), but the species is not known to breed there. The Rocky Mountain Creeper is the form that would be expected to occur, and the curious fact of its absence is emphasized by its known occurrence far to the northwest in Alaska. The Brown Creeper is one of a considerable list of birds that is most unaccountably absent from this region.

Solitaire, Thrushes, and Bluebird

The group of thrushes supplies especially satisfactory examples of strongly differentiated but complementary subspecies on the coast and inland. The Olive-backed and Russet-backed thrushes and the Hermit Thrushes to a marked degree, and the Robins to a lesser extent, are excellent examples of faunal variation. The Olive-backed Thrush ranges from the Atlantic coast northwestward into British Columbia and Alaska, so far as to leave only the narrowest possible coastal habitat for its Russet-backed relative of the Sitkan district. It was found breeding along the Stikine to within forty miles of the

Solitaire, Thrushes, and Bluebird.

Coast	Interior
"Russet-backed" Swainson's Thrush	"Olive-backed" Swainson's Thrush
"Dwarf" Hermit Thrush	"Alaskan" Hermit Thrush
	Gray-cheeked Thrush
"Northwest Coast" American Robin	"Eastern" American Robin
"Coast" Varied Thrush	"Northern" Varied Thrush
	Townsend's Solitaire
	"Arctic" Mountain Bluebird

coast, and, in further evidence of aggressive virility, in the upper Skeena Valley, where so many mainland subspecies show distinct traces of characters of their coastal relatives, and where several coastal species have penetrated in some numbers, the Olive-backed Thrush retains its subspecific character unchanged. This seems to be one of the inland forms that is pressing toward the coast wherever an opening affords. The Russet-backed Thrush, on the other hand, abundant enough in its restricted coastal habitat, shows no inclination whatever to spread inland over the paths that one or two of its associates have followed. Olive-backed and Russet-backed thrushes both are lowland birds. The Olive-backed is characteristic of the quaking aspen woods, of the willow thickets as second choice; the Russet-backed occupies the restricted marginal areas of alder and willow that border the dense Sitkan forests.

Of the Hermit thrushes, the Alaska Hermit inland and the Dwarf Hermit coastwise replace one another faunally just as do the Olive-backed and Russet-backed, with this additional inter-relation, that while the two last-mentioned are occupants of deciduous woods the Hermit Thrushes are primarily of the coniferous forests. The two inland species differ markedly in migration routes. The Olive-backed Thrush travels southward east of the Rockies, down the Mississippi Valley and to a distant Central American winter home, while the Alaska Hermit Thrush is one of few birds of the interior that migrates directly southward to pass the winter in California and other western states. From the Sitkan district the Russet-backed thrush follows the coast southward to western Mexico, the Dwarf Hermit Thrush to the relatively nearby coast of California.

Among migrating Hermit Thrushes passing southward through the Cassiar and Omineca districts, a few birds appear that are closely similar to the Eastern Hermit Thrush (*Hylocichla guttata pallasii*), with bright rufous back and tawny sides, others that are about intermediate between Eastern and Alaska subspecies. These presumably come from some nesting ground to the northeast, a meeting ground of the two forms.

The Gray-cheeked Thrush is on our list through the capture of two or three individuals in the Cassiar district. This bird in its astounding migrations traverses not only the diagonal length of North America from southeast to northwest, but some individuals even pass beyond Alaska into Siberia. This migration route barely includes the northwestern corner of British Columbia.

The Robin of the interior is not the Western Robin, as in southern British Columbia, but is the eastern bird. Its migration route, as with so many species of this section, must lie toward the southeast, crossing the Rockies somewhere north of central British Columbia. In the coastal region there is another subspecies, the Northwest Coast Robin, that may well have been of direct and relatively recent derivation from the Eastern Robin of the interior.

In the wide habitat of the Varied Thrush over western North America, two fairly well differentiated subspecies may be distinguished, one (*naevius*) in a narrow coastal strip from northern California to southern Alaska, the other (*meruloides*) interiorly from Montana to northern Alaska. It is a rather remarkable fact that this bird appears to be absent as a breeding species from the Atlin and Telegraph Creek regions, although the interior form might be expected to occur there and does, in fact, pass through commonly in migration. Throughout the Sitkan district the coast Varied Thrush is a common and characteristic species.

In the northwestern interior Townsend's Solitaire is of fairly general distribution, though nowhere common. This is a bird that has developed no peculiar local races in its extensive habitat in western North America, possibly through a tendency to avoid those regions that present climatic extremes of one sort or another, just as it stops short of the boundaries of the extremely humid Sitkan district. It has, however, been taken there once or twice on migration, as at the mouth of the Stikine, occurrences of straggling individuals that give rise to speculative wonder as to whether they were timidly trying out an uncompromising adventure, or were utterly lost and on the wrong tack.

The Arctic Bluebird justifies its name in pushing northward through the interior of British Columbia and beyond into Yukon Territory and Alaska. It affords a bright bit of color that is always welcome, and it has, too, a pleasing predilection for human society. This is a species that sends migrating contingents down the Taku and Stikine rivers with some regularity in the fall, and it would not be surprising to find it nesting eventually in congenial spots at the mouths of those streams.

Kinglets

As a species the Golden-crowned Kinglet ranges across the breadth of the continent, but it is not as variable as the Ruby-crowned, and the two North American subspecies, Eastern and Western, are very faintly differentiated. In the Sitkan district the Western Golden-crowned Kinglet is of general distribution and of fairly common occurrence, as birds go in that region. It is, in fact, one of the characteristic species of the southeastern Alaska forest. Inland it is extremely rare and restricted to the higher altitudes. I have seen it there, in fact, only in places where the birds might be regarded as an overflow from the coastal region, on or near the east face of the Coast Range.

In contrast, the Ruby-crowned Kinglet is found in fair abundance in both regions, as two rather strongly contrasted subspecies. The Sitkan Kinglet, a dark-colored coastal form, the Eastern Ruby-crown, of paler coloration, each remains absolutely upon its own side of the dividing mountain range. As with the Golden-crown, the Ruby-crown in the Sitkan district is of general distribution, inland breeding mostly at high altitudes, and in both regions occurring in greatest abundance at the beginning of the southward migration.

Kinglets.

Coast	Interior
"Western" Golden-crowned Kinglet	
"Sitka" Ruby-crowned Kinglet	"Eastern" Ruby-crowned Kinglet

Waxwings, Shrike, and Vireos

The Bohemian Waxwing is Boreal, the Cedar Waxwing Temperate Zone, in their summer distribution, and northern British Columbia is one of the few places where their ranges nearly adjoin. The American subspecies of the Bohemian Waxwing is a slightly differentiated (paler-colored) variety of a circumpolar species; in America its breeding ground is apparently entirely in the northwest, though it sweeps southeastward in its winter travels. It seems curious that despite its obvious Asiatic connections, this species sedulously avoids the Pacific slope forests and remains east of the Coast Range. There is more than a suggestion here of the forces that have permitted the populating of the coastal district from southern sources, to the debarring of immigrants from the north. The Bohemian Waxwing breeds commonly as far south as the

Waxwings, Shrike, and Vireos.

Coast	Interior
	Bohemian Waxwing
	Cedar Waxwing
	Northern Shrike
	Red-eyed Vireo
	Warbling Vireo

Stikine Valley, probably somewhat farther south, as molting birds appear in the upper Skeena Valley before the end of the summer.

The Cedar Waxwing nests in interior British Columbia as far north as the Skeena Valley. Its occurrence in Alaska rests upon one July-collected bird from the lower Chickamin River, a locality that yielded various other species of the interior. The Cedar Waxwing is not quite so intolerant of coastal conditions as is the Bohemian and this one occurrence may perhaps be indicative of a future trend in distribution. So far, Bohemian and Cedar waxwings have not been found nesting in the same locality but it is not unlikely that they will be. Both are notoriously erratic in the direction and extent of their migrations and probably irregular, year to year, in the choice of breeding grounds.

The Northern Shrike presents certain parallels to the Bohemian Waxwing. It, too, is probably a New World representative of an Old World species (though not generally recognized as such), and of very much the same general summer distribution as the waxwing, though probably rather more extensively toward the east. While apparently similar in origin and mode of dispersal, the Northern Shrike suggests a possible greater antiquity in its arrival in America, in its slightly more marked differentiation from the parent stock and in its wider distribution in its newer home. As with the waxwing, the shrike has a related species farther south in North America, but with the shrikes the gap is far wider between the northern and the southern species. The Northern Shrike nests southward barely below the northern boundary of British Columbia, in the Atlin region. It avoids the Pacific slope absolutely during the breeding season, and only a few individuals wander out to the Alaska coast at other times.

The Red-eyed Vireo, primarily of eastern North America, ranges northwestward across southern British Columbia even to Vancouver Island, and northward in the interior to the Skeena River (at Hazelton). This apparently is its extreme limit in that direction, for although I saw several thereabout, twenty miles to the northward in Kispiox Valley, in similar surroundings, none was found. The Western Warbling Vireo is in a different category in that it is the western representative of a widespread species, instead of the eastern, as is usually the case in the northwestern interior, extending northward over a wide front. It is common in the Skeena Valley, and, farther north, much less numerous in the Stikine Valley and decidedly rare in the Atlin region. It is unknown upon the Alaska coast.

Wood Warblers

The coastal region is notably poor in its representation of warblers, both as regards species and individuals, while the inland woods, in contrast, have a decidedly abundant warbler population. We naturally associate these birds with sunshine and brightness, so it is no surprise to find the dark Alaska forests so nearly destitute of them.

Of the four warblers that nest in fair numbers in the Sitkan district, two (Lutescent and Alaska Yellow) are coastal variants of widespread species, and two (Townsend's and Pileolated) are western forms that occur unchanged on both sides of the coast range. The Lutescent Warbler is a strongly marked subspecies that ranges from Alaska south to southern California, west of the Coast Range in the north, west from the Sierra Nevada in California. East of the Coast Range it is replaced by the Orange-crowned Warbler and the Western Orange-crown, neither of which the exact distributional limits are known. The Alaska Yellow Warbler is a slightly differentiated subspecies of limited distribution along the coast of Alaska and northern British Columbia. East of the Coast Range it is replaced by the Eastern Yellow Warbler (at least these birds resemble the eastern form more nearly than any other), which ranges throughout the interior of Alaska and northern British Columbia. The coast-inhabiting Alaska Yellow Warbler ascends the Skeena Valley at least to the Hazelton region, where, however, the Yellow Warbler population is of a betwixt-and-between nature, as between the Alaska and Eastern subspecies.

Townsend's Warbler is a western species that at the northern extreme of its range in southern Alaska is almost exclusively coastal. It can be found in the Atlin and Telegraph Creek regions, but only by close search in certain limited types of surroundings; farther south it extends eastward, even as far as Montana. The Pileolated Warbler, a western subspecies of a wide-ranging North American bird, occurs both inland and on the coast. It is mostly on the mountaintops in the interior; in the Sitkan district it is most common along the mainland coast and on the inner islands, progressively rarer toward the outermost islands.

Wood Warblers.

Coast	Interior
"Lutescent" Orange-crowned Warbler	"Eastern" Orange-crowned Warbler
	"Western" Orange-crowned Warbler
	Tennessee Warbler
"Alaska" Yellow Warbler	"Eastern" Yellow Warbler
	"Hoover's" / "Alaska" / "Myrtle" Yellow-rumped Warbler
	"Audubon's" Yellow-rumped Warbler
	Magnolia Warbler
	Blackpoll Warbler
Townsend's Warbler	Townsend's Warbler
	"Grinnell's" Northern Waterthrush
	MacGillivray's ("Tolmie") Warbler
	"Western" Common Yellowthroat
Wilson's ("Pileolated") Warbler	Wilson's ("Pileolated") Warbler
	American Redstart

The Tennessee Warbler, Magnolia Warbler, Blackpoll Warbler, and Redstart are all eastern species that range west and north for varying distances, the Magnolia to the upper Skeena Valley, the Redstart commonly to the Stikine Valley, in small numbers to Atlin, the Tennessee and Blackpoll northward beyond the British Columbia boundaries. They are among the many birds that distinctly characterize the northwestern interior; none has been taken within the Sitkan district. I have seen Redstarts in the late summer on the Stikine at Great Glacier, just above the British Columbia–Alaska boundary, and it is a species that should be looked for as a straggler into Alaska territory at the mouth of that stream.

Hoover's Warbler (a faintly characterized subspecies of the widespread Myrtle Warbler) and Audubon's Warbler are complementary in their distribution, but both characteristic of the country east of the Coast Range. Audubon's Warbler, in southern British Columbia ranging across the mainland and over Vancouver Island, retreats from the coast north of that latitude. It is abundant in the Omineca district, the northern boundary of its habitat lying probably from twenty to fifty miles north of Kispiox Valley. Hoover's Warbler is abundant about Atlin, much less numerous about Telegraph Creek; its southern breeding limits may lie about midway between the Stikine and Skeena rivers. It probably nests along the Stikine almost to the International Boundary, and it is one of the species that might be expected eventually to establish itself upon the coast. This is an aggressively successful bird, its "niche" in the Sitkan district is assuredly unoccupied, and there are places therein that do not seem unsuited to it. It already occurs as a transient, mostly as immatures, descending the big rivers in the fall, then southward along the mainland coast; it is a decidedly uncommon spring migrant, and then only on the mainland and on the innermost islands.

Grinnell's Water-thrush occurs throughout most of the interior of British Columbia. I have seen it at nearly every point I have visited in the upper Skeena and Stikine valleys and at Atlin, but always as a rarity, requiring careful search. It ranges much farther north, but shuns the coast, never having been taken in the Sitkan district.

Tolmie's Warbler practically duplicates Audubon's Warbler in its distribution in the northwest: across southern British Columbia and over Vancouver Island, back from the coast from there northward, common in the Omineca district, and rare in the Telegraph Creek region, its northern limit. This appears to be distinctly a species that is extending its habitat, pushing westward where favorable openings appear. Breeding colonies are established at various points on the southern mainland of the Sitkan district, at the head of Boca de Quadra Inlet, on the lower Chickamin River, and there are adjacent islands where the species might be expected to appear.

The Western Yellowthroat has essentially the same distribution as the Tolmie Warbler in southern and central British Columbia, and it, too, has thrust westward an extension of its summer habitat along the Chickamin River to saltwater. Although an abundant bird in the Omineca district, it is apparently absent from the Stikine Valley, which is puzzling, as I did find it migrating not uncommonly one September along the lower Taku River, Alaska, which is still farther north. It is a rare migrant in the Atlin region, from where the Taku drainage descends. Somewhere to the eastward of Atlin there must be other nesting grounds of the Western Yellowthroat.

Rusty Blackbird

The common northern representative of the blackbird tribe is the Rusty Blackbird. This is another widespread species that ranges *almost* across the continent, blocked off, as are so many others, from the last few miles between the Coast Range and the sea. It is found throughout the lowlands of the Cassiar and Omineca districts, and breeds along the Stikine to within a hundred miles of the coast, but it is only an occasional straggler that wanders down to the river's mouth at the end of the summer. The southward migration of the Rusty Blackbird is noticeably to the eastward, east of the Rocky Mountains. Very few stray directly south into southern British Columbia.

Louisiana Tanager

Tanagers are primarily tropical or sub-tropical in habitat, and only one species extends up into the Canadian northwest. The Louisiana Tanager is common in southern British Columbia, it extends northward in the interior in some abundance as far as the Omineca district, about Hazelton, and its normal northern limit is apparently the upper Stikine Valley, where it is uncommon. There is a record of a specimen collected on the Alaska coast, at the mouth of the Chickamin River, where it was apparently nesting, a place where other inland birds also were found.

Finches and Sparrows

An outstanding feature of the comparative lists of finches and sparrows is the far greater number of species in the interior. There is just one bird, the Pine Siskin, that is found in abundance, and unchanged in appearance, in both regions. Throughout its wide habitat the Pine Siskin pays some attention to zonal limits (it is mostly of the Canadian and Hudsonian life zones), but the boundaries of faunal areas are commonly ignored, here as elsewhere. There is no fringilline species characteristic of the interior that breeds also upon the coast. One species that is primarily of the coast, the Rusty Song Sparrow, is widespread over the southern half of British Columbia. Even as far north as the Skeena River it extends far inland, and in abundance. On the next large stream to the north, the Stikine, the Song Sparrow numbers are greatly lessened, and, although it occurs upstream as far as Telegraph Creek, it is only in small and scattered colonies at exceptionally favorable localities. This is about the northern limit of the species inland, though along the coast Song sparrows, in one form or another, extend much farther, though west rather than north, even to the Aleutian Islands.

There are two finches that in small numbers extend northward this far from more southern centers of abundance. The Evening Grosbeak, common in places in the southern interior of British Columbia, ranges north to the vicinity of Hazelton. It has never reached the Alaska coast. The Eastern Purple Finch (extending westward from the Atlantic Coast) is not uncommon in the Hazelton region (the Omineca district), it is decidedly rare in the upper Stikine Valley, and little more than a straggler to Atlin, the northernmost point at which it has been taken. It is worth noting that the western subspecies of the Purple Finch, the California Purple Finch, is found on Vancouver Island and the adjacent mainland coast, but that in British Columbia the boundaries of these two subspecies are nowhere known to approach within a hundred miles of each other.

Of the crossbills, the small American Crossbill is common in the Sitkan district; the large Bendire's Crossbill (not common) has been found about Atlin and Telegraph Creek, and undoubtedly occurs, though probably erratically, as crossbills do, from those points southward throughout the interior. There is a curious quirk in distribution here, in that the bird of the Sitkan district (it occurs south into California along the coast) is indistinguishable from the common crossbill of eastern North America, though the habitat of Bendire's Crossbill intervenes between the two.

The White-winged Crossbill is the common species of the interior. It migrates in winter into the Sitkan district, and I believe that, in some seasons, at least, it may nest along the mainland coast. I have found it obviously doing so, though no nests were discovered, along the Stikine within thirty miles of the river's mouth. As a point in local distribution, the Red Crossbill of the interior (Bendire's Crossbill) occupies the stands of jack pine, while the White-winged Crossbill of the same region is just as rigidly confined to the white spruce. The Red Crossbill of the coast (the American Crossbill) has no extensive jack pine woods to resort to, and cheerfully occupies the spruce forest.

It seems curious that the Red Crossbill should be of more southern distribution than the White-winged, for our several American subspecies of the Red Crossbill are but slightly differentiated variants of a species that extends also across Northern Asia and Europe, and they, therefore, might be expected to range as far north in the continent as could any member of the genus.

The Common Redpoll, transcontinental in habitat, is an inland species, probably altogether of the interior during the nesting season. It is known to nest only as far south as the Atlin region. Occurrences in the Sitkan district are merely of transient flocks.

There is another coastal species, the Kadiak Pine Grosbeak, that has been found in summer in the upper Stikine Valley, possibly a parallel occurrence

Finches and Sparrows.

Coast	Interior
	Evening Grosbeak
"Kadiak" Pine Grosbeak (*Pinicola enucleator flammula*)	"Kadiak" Pine Grosbeak
	"Alaskan" Pine Grosbeak (*P. e. alascensis*)
	Purple Finch
"American" Red Crossbill (*Loxia curvirostra minor*)	"Bendire's" Red Crossbill (*L. c. bendirei*)
	White-winged Crossbill
"Hepburn's" Gray-crowned Rosy-Finch	
	Common Redpoll
Pine Siskin	Pine Siskin
"Kadiak" Savannah Sparrow (*Passerculus sandwichensis anthinus*)	"Western" Savannah Sparrow (*P. s. alaudinus*)
	"Gambel's" White-crowned Sparrow
	Golden-crowned Sparrow
	White-throated Sparrow
	Tree Sparrow
	Chipping Sparrow
	"Timberline" Brewer's Sparrow (*Spizella taverneri*)
"Oregon" Dark-eyed Junco (*Junco hyemalis oreganus*)	"Shufeldt's" Dark-eyed Junco (*J. h. shufeldti*)
"Cassiar" or "Slate-colored" Dark-eyed Junco (*J. h. cismontanus*)	
"Rusty" Song Sparrow (*Melospiza melodia morphna*)	"Rusty" Song Sparrow

Coast	Interior
"Sooty" Song Sparrow (*M. m. rufina*)	
"Forbush's" Lincoln Sparrow (*Melospiza lincolnii gracilis*)	Lincoln Sparrow (*Melospiza l. lincolnii*)
"Townsend's" Fox Sparrow (*Passerella iliaca townsendi*)	"Alberta" Fox Sparrow (*P. i. altivagans*)
"Sooty" Fox Sparrow (*P. i. fuliginosa*)	

to the Song Sparrow. Distributions and manner of occurrence of the several subspecies of the Pine Grosbeak in the northwest are imperfectly known but it looks as though there were a coastal form, not at all common in the Sitkan district, that may stretch inland for some distance up such valleys as the Stikine. The Alaskan Pine Grosbeak (the subspecies of the northern interior) is not known to breed even as far south as the Atlin region.

Different subspecies of the Savannah Sparrow replace one another most satisfactorily inland and on the coast; the gray-colored, slender-billed Western Savannah Sparrow throughout the interior, the brown-colored, stubby-billed Kadiak Savannah Sparrow in the Sitkan district. Throughout the interior of North America breeding Savannah Sparrows are in continuous, or nearly continuous distribution from about latitude 40 degrees northward throughout the lowlands. The bird of interior northern British Columbia, Yukon, and Alaska is a northern offshoot of this assemblage, and may be assumed to have reached its present home by direct immigration northward when conditions permitted. Another direct offshoot of this same inland stock has reached the coast in extreme southern British Columbia, at the mouth of the Fraser River, but so far as is known there are no Savannah Sparrows breeding along the coast between that point and Dixon Entrance. The Western Savannah Sparrow migrates southward and southeastward, one of very few species of the region that moves even in part into California during the winter. Its winter home extends from the coast of southern California east and into western Texas.

The Kadiak Savannah Sparrow, with summer habitat along the coast from Kodiak Island southeastward into the Sitkan district, I believe to be of rather remote relationship to the bird of the interior. I believe it to be derived from the large Aleutian Savannah Sparrow (*P. s. sandwichensis*), and an invader that pushed its way southward as the glaciers withdrew; it is a relatively recent arrival in the Sitkan district and one that even now has only partly occupied that territory. The Kadiak Savannah Sparrow moves directly southward, hardly beyond the coast of northern California.

Similarly, Lincoln's Sparrow (inland) and Forbush's Sparrow (coastal) are complementary in their distribution. Lincoln's Sparrow extends uninterruptedly across subarctic North America, as one uniform subspecies from the Atlantic west to the Coast Range. It is a fairly common bird in the interior of British Columbia. In the Sitkan district the slightly differentiated subspecies, Forbush's Sparrow, is present in abundance, probably in as great numbers as occur anywhere within the widespread range of the species. Forbush's Sparrow is, I believe, directly derived from the stock of the adjacent interior, and not an immigrant from the southern coast. It is known to occur along one connecting valley, the Stikine, and probably does so along others as well. This is not a very strongly marked subspecies, but at that it has responded to environmental influence more than most Sitkan races of similar origin. Lincoln's Sparrow of the interior migrates southward far into Mexico. Forbush's Sparrow travels southward along the coast, for the most part no farther than central California.

In the group of Juncos we are here at the meeting point of three distinct forms, Cassiar Junco, Shufeldt's Junco, and Oregon Junco. The Slate-colored Junco, the prevalent form of eastern and northern North America, occurs over most of Alaska, north of our region, but in the Atlin and Stikine territories it is represented by a slightly different subspecies, the Cassiar Junco, a rather curious variant. In this bird the male is closely similar to the male Slate-colored, to the northward, while the female is closely similar to the female Shufeldt's Junco, to the southward. In their southern winter habitat male and female of the Cassiar Junco without doubt have been many times recorded as, respectively, of those two species. The Cassiar Junco, part of the Slate-colored aggregation, represents an extreme western outpost of that widespread species. Shufeldt's Junco, of interior British Columbia north to the Skeena Valley, and the Oregon Junco, of the Sitkan district, are closely related subspecies of the Black-headed Oregon Junco aggregation, which occupies the Pacific slope from southeastern Alaska south into northern Lower California. At the meeting point of Shufeldt's and Oregon Junco in the Skeena Valley, of Shufeldt's and Cassiar Junco a little farther north, and of Cassiar and Oregon Junco on the lower Stikine, there are produced some birds that are curiously and variously "intermediate" in their characters of color and markings, fearfully puzzling things to the systematist when they reach his hands from some far southern winter resort, with no clue as to their place of origin.

In the Fox Sparrows (*Passerella*) conditions are the reverse to those in the Juncos, in that there is just one subspecies inland, the paler-colored Alberta

Fox Sparrow, and two subspecies in different parts of the coast, the extremely dark-colored Sooty and Townsend Fox Sparrows. The Alberta Fox Sparrow occurs northward to the Skeena drainage, where it occupies the Hudsonian zone on the higher mountains. Curiously, there is a gap throughout the Stikine and Atlin countries (the Cassiar region) where apparently no form of Fox Sparrow occurs, a wide gap between the ranges of the Alberta Fox Sparrow and the boreal Eastern Fox Sparrow, which is found from the Atlantic coast westward.

Hepburn's Rosy Finch is a coastal variety of a species that is widespread over the higher mountains of the northwest, and probably occurs wherever conditions are favorable above timberline throughout the Sitkan district, on islands and mainland. It has been found in summer on the eastern face of the Coast Range in the upper reaches of the Stikine, and on a mountain near Hazelton, on the upper Skeena River. The Gray-crowned Rosy Finch, its analogue in the interior, although known from localities to the southeast and to the northwest, has not been found in the region we are here considering.

The White-throated Sparrow is primarily of eastern North America, but has been found in the vicinity of Hazelton, apparently its extreme northwestern limit. The other crown sparrows, Gambel's and the Golden-crowned, are characteristic and abundant species in the interior of the northwest, Gambel's in the lowlands, the Golden-crowned at high altitudes. Gambel's Sparrow frequents the thickets of willow, wild rose, or the various berry bushes, at the edges of the woodlands, the Golden-crowned the willow-bordered brooks and the balsam thickets above timberline. Both are migrants through the Sitkan district, in fair abundance along the mainland coast, rather rare on the outer islands. Both of these birds are peculiar in their occurrence in the interior of the northwest, especially so the Golden-crowned, in that they are western species, primarily of the Pacific slope, and in their summer habitat they are part of a population composed mostly of eastern migrants and boreal residents. There are not many exceptions to the general rule that Pacific slope species in far northern migration remain west of the Coast Range.

The Western Tree Sparrow, subarctic in habitat and, as a species, extending across the continent, finds its southern breeding limit in the Cassiar district. It is not known to nest south of the Stikine. In the Atlin and Stikine sections it is a timberline bird, haunting the willow thickets along brooks and lakelets above the forest level. If it reaches the coast at all it is as a scarce migrant. The closely related Chipping Sparrow, common throughout the interior, is the Eastern variety, and not the Western Chipping Sparrow found west of the Rockies farther south. It is a lowland bird, does not cross the Coast Range anywhere, and in southward migration swings east of the Rocky Mountains before the United States boundary.

A recent (1924) addition to the North American bird list is the Timberline Sparrow, a *Spizella* that is closely similar to Brewer's Sparrow of the sage-brush deserts of the Great Basin. The Timberline Sparrow has thus far been found in summer only in the Atlin region, where it occurs at relatively high elevations, in the regions occupied also by Rock Ptarmigan, Golden-crowned Sparrow, and Horned Lark. It is mostly at a higher level than the Tree Sparrow, of the same genus, and has not been detected in the lowlands even in migration. Nothing is known of its distribution in winter, and only one or two migrants have been collected, in southern British Columbia and in Montana.

SECTION 2

The Story behind the Research

(Field Journal Entries, Maps, and Expedition Itineraries)

An ornithologist's field journal is similar to a diary, except that while a diary is often composed of personal feelings, a field journal is filled with natural history observations and scientific data. Everything of biological relevance that happens each day is recorded in the field journal, and, for an ornithologist, this would include the weather, habitats visited, routes traveled, and most importantly—a species list. Behavioral observations, ideas, and theories may also be written down, and hand-drawn maps and sketches are sometimes added. The success of an expedition could depend as much on the records and notes in the field journal as on the number of specimens returned to the museum.

A field journal is kept using a standard format, making it straightforward to complete each day's entries and easy to retrieve information later on. Swarth used the Grinnell System to keep his field journal. Joseph Grinnell, the first director of UC Berkeley's Museum of Vertebrate Zoology (MVZ), refined and promoted this technique. In 1910, Grinnell wrote, "The field collector is supplied with a separate-leaf notebook. He writes his records on the day of observation with carbon ink, on one side of the paper only. The floral surroundings are recorded, especially with respect to their bearing on the animal secured. The behavior of the animal is described and everything else which is thought by the collector to be of use in the study of the species is put on record at the time the observations are made in the field. These field notes are filed so as to be as readily accessible to the student in the museum as are the specimens themselves." Grinnell required all MVZ researchers to use this field journal format (for more on the Grinnell System, see Remsen 1977; Herman 1986). Swarth's field journals, and those of many other MVZ scientists, are archived at the museum, and many have been digitized for viewing online (https://ecoreader.berkeley.edu/). Field journal entries were accessed from the MVZ EcoReader: Swarth, H. S., Field Notebook, vols. 1679 (section 1), 1680 (section 1), 1683 (section 2), 1684 (section 2), and 1685 (section 2).

Swarth '09   Etolin Island   107

Birds are scarce, and, from the continuous raining and stormy weather, doubly hard to find. Of water birds there are: Am. Mergansers, fairly abundant; one brood of young seen; Pigeon Guillemot, quite abundant out in the channel; Glaucous-winged Gull, one or two; Am. Scoter; Loon; Spotted Sandpiper (1); Greater Yellowlegs (1).

Land birds :- Hermit Thrush; Varied Thrush (one or two heard, not seen); Winter Wren; Western Flycatcher; Song Sparrow (one pair with young); Junco; Steller Jay; Raven — a brood of young with their parents has been staying around the cabin and eating the refuse thrown out; they are extremely noisy, and make the most horrible gurgling and croaking noises; Beach Crow — common, many young ones out flying around; I saw them here, for the first time, getting clams and flying in the air with them in order to drop them on the rocks and break them. They were doing it right along. Golden-crowned Kinglet, young and old; Harris Woodpecker; Red-breasted Sapsucker, there is a brood of young in a stub not far from camp; Grouse, apparently not abundant is but few were heard hooting; Flicker, I heard one several times but never caught sight of it; Red Crossbill, a flock was seen flying overhead nearly every day, but I never saw them alight; Bald Eagle, abundant.

We are camped in a bay at the south end of the island, at an abandoned fish camp, where a small stream empties out. The country in the immediate vicinity is flat, with the hills some distance back; along the beach is a strip of heavy timber, spruce, hemlock, and a good deal of cedar, with considerable underbrush, two or three hundred yards wide. In back of that are long stretches of parks.

A page from the field journal. Etolin Island, Alexander Archipelago, southeast Alaska, 11 July 1909. Accessed from EcoReader (Swarth, H. S., Field Notebook, vol. 1679), Museum of Vertebrate Zoology Archives, University of California, Berkeley.

1909 Field Journal Entries

Waters off Kupreanof Island, Alaska; 10 April 1909

All afternoon, as long as we were in deep water, scores of porpoises were playing about the boat. Keeping close alongside, or just in front, there was hardly a minute when there were not some in sight. They would apparently spot us from a distance and come headlong, like a lot of boys playing leapfrog. They would escort us a mile or two, and then some others would appear and the first lot would draw off. We could plainly hear them gasp when they reached the surface and see them dispel the air they contained just before coming to the top.

Kuiu Island, Alaska; 28 April 1909

In the morning, while I was watching a flock of ducks, several mergansers came quite close, and they were feeding in a manner that was new to me. I have usually seen them diving for their food, but these swam along, holding their bills close to the surface of the water, at a slight angle so that about half the head was submerged, in much the same way that a shoveler feeds. I saw both species, the Red-breasted and the American, doing this.

The channel between Prince of Wales and Suemez Islands, southeastern Alaska, 1909. Photograph by Harry S. Swarth, with permission of the Museum of Vertebrate Zoology, University of California, Berkeley.

Heceta Island, Alaska; 24 May 1909

In several places we passed through large, scattered, companies of the Pacific Loon—at any rate smaller than *immer* and with a gray head. Other waterbirds seen were cormorants, White-winged Scoters, and Marbled Murrelets, but not many of either. Eagles were more numerous than at any place thus far—we passed one rocky little islet where they seemed to be after the spawning herring and there must have been forty or fifty at least, in the space of about an acre.

Dall Island, Alaska; 4 June 1909

Left "Rocky Bay" about 8 a.m. and coasted down the west side of Dall Island. The weather was good enough so that we might have gone out to Forrester Island, but it was necessary to know some harbor that could be run into in a hurry if need be. There are several such on the chart, but they have no existence in reality! We finally came to one directly opposite Forrester Island, sheltered on all sides, and all that could be desired. The cove we are in is an excellent place for shelter, but a wretched place for birds; I saw none at all.

Duke Island, Alaska; 8 June 1909

I put in most of the day skinning the accumulation of Ancient Murrelets of the last two days. There is quite a little variation in plumage among them, regardless of sex, mainly in the admixture of gray in the black throat. Some of the females had laid their sets, and were incubating, judging from the bare spots on the abdomen and the condition of the oviduct, but one at least had not. The male bird secured had the two denuded spots on the belly also, so I suppose he does his share of the incubating also. Yesterday I had one or two good opportunities of watching them close under the bow of the launch, and could plainly see how they used their wings under the water, practically flying, though with their last joint, the "hand," almost parallel with the body and not outstretched, as when going through the air.

Boca de Quadra, Alaska; 12 June 1909

This morning I went to the place where the swifts were yesterday, and prepared to camp there as long as my ammunition lasted. They were there again, but more of them, and for the most part flying high out of reach. The Black Swifts fly much more slowly than the little Vaux, with more soaring and much less fluttering of wings; while they were not chasing each other about at all, as the smaller species was doing all the time. Also, they were quiet, while the others were twittering all the time.

Portage Cove slough, Revillagigedo Island, southeastern Alaska, 1909. Photograph by Harry S. Swarth, with permission of the Museum of Vertebrate Zoology, University of California, Berkeley.

Boca de Quadra, Alaska; 13 June 1909

Of the birds, the most interesting to me, was finding the two species of swifts here. One of the Black Swifts contained an egg about half formed so I suppose it is fair to suppose that they are breeding, or will do so, somewhere near here. The prospectors told Hasselborg that these swifts had just arrived here a day or two before they started back into the mountains, say about ten days ago, and they had noticed them flying over the meadow. This must be about their northern limit.

Chickamin River, Alaska; 22 June 1909

It is of some interest to note that all of the more uncommon birds I have been getting lately—Yellowthroat, Cedar bird, Traill Flycatcher, Wood Pewee, Tanager, and Yellow Warbler, my attention has been called to their presence, in every case, by the call note or the song, sometimes so faintly heard that it took long persistent search to locate the author; nor would I have got any of them in any other way, for this is a very hard place to see birds in. It illustrates what a great handicap deafness is to a collector.

Field assistant Allen Hasselborg on Admiralty Island, southeastern Alaska. Photograph by William L. Finley. Courtesy of the Oregon Historical Society.

Chickamin River, Alaska; 27 June 1909

Song Sparrows are quite abundant in the meadows, and undoubtedly with complete sets of eggs—some possibly with young. Forbush's Sparrow also is equally numerous and also incubating.—Two species of the same genus inhabiting the same place precisely, and in about equal numbers. There are some Savannah Sparrows breeding in the meadows also, but not very many.

Revillagigedo Island, Alaska; 2 July 1909

Yesterday Hasselborg found a Flicker's nest, and shot the female bird, and today we went together to the spot. He pounded on the stub, and I shot the male bird as it flew out. It is a "hybrid" with the head markings of *auratus*, and the red wings and tail of *cafer*. The nest was in a dead stub about fifty feet from the ground—so rotten that by climbing up a few feet Hasselborg managed to shake the whole top off it, the part containing the nest. It was just rotten punk, and simply disintegrated when it came down. In the debris we found the remains of five young birds just out of the shell, and one rotten egg. The nest was in the valley close to the river—the timber there is very scattering, with considerable short underbrush and many dead trees; quite open country such as seems to suit Flickers requirements.

Wrangell Island, Alaska; 16 July 1909
Raining in torrents all last night, and all day today, as in fact it has been doing ever since we struck this place. Also I have been laid up for several days, either from spoiled milk, spoiled meat, or moldy bread—we have all three—and have been unable to do anything at all. This would be a wretched place in the best of weather, and under the circumstances there is literally nothing doing.

Mitkof Island, Alaska; 8 August 1909
This was a red-letter day, for it did not rain at all, in fact the sun was out and shining nearly all day long. It was cloudy around the edges of course, but overhead it was clear blue sky! Last night I heard a wolf howling somewhere nearby, and this morning there were fresh tracks less than a quarter mile on either side of camp, on the beach.

Taku River, southeastern Alaska, 1909. Photograph by Harry S. Swarth, with permission of the Museum of Vertebrate Zoology, University of California, Berkeley.

Hasselborg's boat beached on a mudflat at low tide, Taku River, southeastern Alaska, 1909.
Photograph by Harry S. Swarth, with permission of the Museum of Vertebrate Zoology,
University of California, Berkeley.

Taku River, Alaska; 4 September 1909

At 3 a.m. we started up the river, with the tide, and almost immediately ran
into quantities of floating ice. Some of this hit the launch pretty hard, and
shortly she started to leak. The first intimation we had of the state of affairs was
derived from the flywheel, which running under water, sent a fountain all over
the cabin. After that I had to bail steadily while Hasselborg ran the boat, and
kept it up until we beached her at a fishing camp.

Taku River, Alaska; 9 September 1909

This duck (mallard) was shot at Port Snettisham, August 31. At the time I had
just time to take the skin off, and have been preparing it by degrees ever since.
It was beastly fat, one of the fattest ducks I have ever tackled, and the skin
was consequently as tender as wet tissue paper. To add to the difficulties it
is one mass of pin feathers. I scraped it all I could—all I dared to—and then
put it to soak in gasoline, where it stayed for two days, with occasional scrap-
ing and squeezing out of grease. After drying it I found it impossible to fluff
out the feathers—being nearly all pin feathers they were so loosely attached.
Consequently, it is a wretched specimen, in appearance much like one of those
wooly haired chickens! A dozen times I was on the verge of throwing it away,
but saved it as it is a plumage not represented in the museum collection.

Taku River, Alaska; 16 September 1909

The Sparrow Hawk I shot today was secured by somewhat of a fluke. I saw two small hawks sitting on neighboring tree tops, and taking them both to be Pigeon Hawks, as one of them was, started after the nearest. This was the bona fide Pigeon Hawk, and I was almost within gunshot when the other left his perch, drove the Pigeon Hawk away, and occupied his place, where I shot him. He lodged in the top of the tree—a spruce some sixty or seventy feet high—and I was much inclined to go off and leave him, as it was raining hard, and it was no joke to climb a tree in gum boots anyway, but I finally went up after him. Not until I had the bird in my hand did I realize what it was.

Taku River, Alaska; 27 September 1909

Spent the day in packing specimens and otherwise preparing to break camp tomorrow. This last collecting ground of the season has in many respects proved the most interesting of any place we have visited. My camp is on the south side of the Taku River, some fifteen miles from the mouth, while Hasselborg has been staying a couple of miles or so further upstream. The river hereabouts is from a half a mile to a mile wide, with steep, abrupt banks all along its course, and no shore at all for waders, which consequently are entirely absent. At this place there is an area some two miles long and a half

Chickamin River, southeastern Alaska, 1909. Photograph by Harry S. Swarth, with permission of the Museum of Vertebrate Zoology, University of California, Berkeley.

Fireweed in bloom, Mendenhall wetlands near Juneau, southeastern Alaska. Photograph by John Schoen.

a mile wide—approximately—a little ways back from the stream, grown up with dry woods such as I have seen nowhere else. There is a scattering growth of cottonwoods with a few birch and a sprinkling of spruce trees, and very little underbrush. The ground is dry and gravelly and there is almost no grass. This is the best place for birds, that is a great number of individuals, that I have seen anywhere in this country. It was an excellent place for collecting, for both birds and mammals are abundant, and I would have got far more than I did if we had only had a little decent weather.

1910 Field Journal Entries

Golden Eagle Mine, Vancouver Island, BC; 14 July 1910

A walk of half a mile brought us to two little lakes, one about an acre in extent, the other about half as large, fed by surrounding snow banks, in places twenty or thirty feet high, like miniature glaciers at the water's edge. We camped in a sheltered thicket at a corner of the smaller lake, and had lunch. While we were sitting, resting, we both were nearly asleep, for we were very tired, but I was startled by a slight splashing in the water just below us. Looking down I saw that a bear had quietly swum across the lake without noticing our camp in the trees, and was just wading out on the bank about thirty yards from us. I whispered to Despard and the bear heard me, and turning caught sight of our dog. The two stood gazing at each other curiously, without a sound, but in the meantime Despard secured his rifle, and just as the bear turned away he fired. It gave one leap into the thicket and disappeared.

Great Central Lake, Vancouver Island, BC; 19 August 1910

On the morning of the 19th Despard and I started on our Ptarmigan hunt. We left about 10 a.m. with light packs, and as the spot we were heading for was said to be eleven miles from the lake, supposed we could reach it early in the afternoon. The trail was fairly good tho much encumbered by fallen trees, but the day was oppressively hot and we did not make good time. Also there were several streams to be crossed, the bridges being sometimes broken, and sometimes entirely gone. At one point a bridge crossed a narrow gorge seventy or eighty feet above the water, which fell in a thunderous cataract, and the bridge was pretty shaky and apparently insecure, tho probably safe enough.

Great Central Lake, Vancouver Island, BC; 20 August 1910

On the morning of the 20th we started on the climb up to the "Big Interior Mine." The trail led to the foot of a cliff, up which one has to scramble as best he may. The first quarter of the climb is about the worst, but that is so bad that

Field camp at Errington, Vancouver Island, British Columbia, 1910. Photograph by Harry S. Swarth, with permission of the Museum of Vertebrate Zoology, University of California, Berkeley.

we left most of our packs behind us. We discarded blankets, camera, and rifle, and went on with just the bare necessities—provisions, shotguns, and ammunition. There were several very bad places to be passed. First the face of the cliff, slippery with dripping water, then a perpendicular rock, on which a rough ladder had been fastened, then some rocky places up which we scrambled with the aid of a rope. Further on was another perpendicular place against which a tree had been felled. At any of these places a slip would have meant a fall of at least sixty or seventy feet, and I was glad when they were passed.

Great Central Lake, Vancouver Island, BC; 26 August 1910

At 6:00 p.m. we started our trip down the lake. We travelled at night as it is usually calm then, and lately there has been a strong breeze blowing up the lake every day. Our canoe was pretty heavily loaded and leaked besides, so we did not want to take any chances in a strong wind. Despard and I took two hour shifts rowing, and we stopped an hour at midnight for supper. The moon came out at that time, so travelling was quite pleasant, especially as the night was not cold enough to be uncomfortable. We reached the foot of the lake at 6:00 a.m.

Mount Saunders, Vancouver Island, British Columbia, 1910. Photograph by Harry S. Swarth, with permission of the Museum of Vertebrate Zoology, University of California, Berkeley.

Field Camp at Beaver Creek, Vancouver Island, British Columbia, 1910. Photograph by Harry S. Swarth, with permission of the Museum of Vertebrate Zoology, University of California, Berkeley.

Mount Arrowsmith, Vancouver Island, BC; 6 September 1910

There were a mile or two of very bad travelling. There is no trail, of course, beyond the mine, and as the mine has not been worked for at least ten years, and the trail neglected ever since, it's nothing to brag about. After getting thru the burnt area we struck green timber with much better going, and about 6:30 fortunately struck a gulch with a little water in it. We camped here, and as the night was clear, and a north breeze blowing, did not put up the tent. The following is what we carried: Despard, a little pup tent, a shotgun, one box shells, tea, sugar, salt, butter, cheese, four tins of meat, two apples, tea pot, and sweater; myself, one pair blankets, rubber tarpaulin, sweater, shotgun, ax, one box shells, hard tack, a cold grouse, two apples, one loaf of bread, four tins of meat, two cups. . . .

. . . The weather was miserable all day, cloudy, with occasional showers, and with a bitterly cold southeast wind blowing. We were both soaked thru, and anything but comfortable. In the evening we built a big fire and got partially dried off, and then went to bed to get warm. It was a wretched night tho, cold and uncomfortable, while the noise of the wind kept us awake most of the time, and it got under the edges of the tent despite all we could do. In the morning we crawled out, stiff with cold, and as the weather was evidently getting worse, decided to give up the hunt, and get down as quickly as we could.

We packed up and started the descent about 8, going down a different ridge from the one we came up. Soon after we started it began to snow quite heavily; we had no compass, and losing our bearings, spent about two hours walking around a little mountain. Then the clouds lifted a little and we saw that we had to make a fresh start. . . . The trip was, of course, a failure as far as Ptarmigan were concerned, and it was an exceedingly arduous and uncomfortable one, but if the weather had been favorable, I believe we would have been successful.

1919 Field Journal Entries

The Junction, Telegraph Creek, BC; 26 May 1919

Juncos are perhaps more numerous than anything else, and one may find a pair of these birds perhaps every three or four hundred yards. I shot three, one of which I failed to find. Dixon shot two more. Most curious looking birds, of the *hyemalis* type rather than *oreganus*, but the two females have the sides distinctly pinkish. The males are slaty, but the head is appreciably darker than the back, and the black throat is cut off straight across the breast rather than merging into the slaty of the sides, as in eastern examples of *hyemalis*. This all has bearing on the relationships existing between *hyemalis* and *oreganus*, but it's too early yet to attempt to read the puzzle.

Stikine River, near Telegraph Creek, British Columbia, 1919. Photograph by Harry S. Swarth, with permission of the Museum of Vertebrate Zoology, University of California, Berkeley.

Joseph Dixon and
Joseph Grinnell, Pilot
Knob, California,
1910. Photograph by
Joseph Dixon, with
permission of the
Museum of Vertebrate
Zoology, University of
California, Berkeley.

The Junction, Telegraph Creek, BC; 5 June 1919

The Golden-crowned Sparrows are apparently on their nesting grounds on the
summit; we do not see them at the lower elevation of our camp at all. The
birds' actions are indicative of their breeding, the males have the testes largely
developed, and in the female taken today the eggs are beginning to form. . . .
It seems curious that the Canada Jays should be molting so extensively at this
time. The one that Dixon shot today was in the very midst of the molt. . . .
Dixon tells me that the peculiar song I heard yesterday, and which I ascribed to
some unidentified warbler, is one song of the Golden-crowned Sparrow; that
he saw the bird give it.

Harry Swarth standing among devil's club (*Oplopanax horridus*), a stout, spiny understory shrub, 1919. Photograph by Joseph Dixon, with permission of the Museum of Vertebrate Zoology, University of California, Berkeley.

Telegraph Creek to Glenora, BC; 26 June 1919

After many days of waiting for the river boat, we finally decided to go by team to Glenora. Our driver is Mr. Cox, of this town, our vehicle a light wagon. It was no small task to get our outfit upon this one rig, but it was finally accomplished, all but one chest, and we started out. Scarcely a mile from town our troubles began. First one of the braces of the tongue broke; this was fixed, but the front end of the wagon continued to disintegrate. The king bolt went and then the tongue gave way entirely, some five miles from town. Mr. Cox announced his intention of making a "jumper," and as neither of us knew what a jumper was, we stood and watched until it began to take shape as something between a sled and a stone boat. As our driver expressed every confidence in his ability to get thru, we piled half our stuff on this contraption, and started. The jumper did not work very well. It came apart at every possible joint, and had to be joined, tied, nailed or wired as the case might be. Cox told us he had a good jumper cached by the road some three miles from Glenora, but a little later we learned that he had placed it there twelve years ago! People here disregard the passage of time in a most extraordinary way. We eventually reached this jumper and transferred our load. All went well for about two miles, when the braces holding the two runners together all gave way at once, the resulting collapse equaling that of the One Hoss Shay. An hour's labor here produced what was virtually a new conveyance, and this time we reached our destination with no more breakdowns. Left Telegraph Creek at 11:30 a.m., reached Glenora at 12:30 a.m. Distance travelled, fourteen miles.

Glenora, BC; 5 July 1919

Today we walked to Telegraph Creek and back, some twelve miles each way. The trip was primarily to examine a waxwing's nest, left over from our previous work there, but we also had various things to attend to in town.

... Left Glenora about 8:30 a.m. and reached Telegraph Creek about 2 p.m. We made the return trip in four hours (reached home at 8:45 p.m.) having but few stops to detain us. We were both pretty tired.

... I have been trying for some time to think of an expression to indicate the difference between the call notes of the Bohemian and Cedar waxwings. Perhaps as good a term as any is to say that the note of the former is more of a *buzz*, while that of the latter is a *hiss*.

... Mosquitoes were at their very worst all day. I never saw a region where they are so uniformly distributed, so abundant, and so ferocious. The only place where there is any respite from their attacks is on some of the exposed points of the bluffs over the river, where the breeze keeps them away. The town of Telegraph Creek is fortunately situated on such a spot, and there are few mosquitoes there. We wore head nets all day as a matter of course, and at Telegraph Creek I bought a pair of cheap cotton gloves as protection for my

Skeena River, British Columbia, 1921. Photograph by Harry S. Swarth, with permission of the Museum of Vertebrate Zoology, University of California, Berkeley.

hands. They sting me repeatedly on the shoulders, where my shirt draws tight on the skin. Today one stung me over the left temple, apparently an especially tender spot, for the whole side of my head swelled up, far worse than from a bee sting and far more painful.

Doch-da-on Creek, BC; 23 July 1919

The *Leucostictes* that were about the snow banks were obviously feeding on insects to some extent. In one case when I was approaching an adult, it suddenly flew into the air and straight toward me, hovering a moment and capturing a flying insect less than twenty feet from where I stood. Nearly all the *Leucostictes* shot had their crops filled, and I saved the contents of all.

Doch-da-on Creek, BC; 25 July 1919

The boat failed to arrive. Ted had told us that they might possibly lay over a day, tho he did not think they would, so we had to get ready any way. Being all packed up to move, there was nothing we could do, and as the day was sultry and the mosquitoes extra pernicious, we took a picnic lunch out to a gravel bar sparsely grown with willows. Mr. and Mrs. Jackson, Doreen Dodd, Dixon and myself, and besides lunch we had a pile of magazines, cards and a cribbage board, and the field glasses. There we spent a delightfully lazy after noon, the first relaxation Dixon and I have had on this trip.

Doch-da-on Creek, BC; 7 August 1919

The Juncos here present a curious medley of characters. Some are exactly like the Telegraph Creek birds, and others are the "*shufeldti*" type. The latter is possibly an indication of hybridism with the coast *oreganus*. There is nothing like gradual intergradation between the two. I have tried to get as many juncos as possible here, but it has been difficult; also so many of those collected are juveniles, and it is doubtful if these will be of much value in comparison of the races.

Sergief Island, Alaska; 23 August 1919

On the way home (I was alone at the time) I saw a large bird circling about high overhead. At first I could think of nothing but a California Condor. It was of enormous size, with broad wings and rather short tail, and plainly showed white patches at the bend of the wing as it wheeled about. All this fits the description of the Kamchatka Sea Eagle, and while this sight identification cannot stand as an absolutely authentic record for this species, I do not see what else it could have been.

1921 Field Journal Entries

Hazelton, BC; 10 June 1921

As we went past the railroad station this morning on our way to the burnt timber, a bird was heard singing in a nearby thicket, a thrasher-like song that puzzled me completely. Strong went after the bird and secured it with little trouble. It proved to be a catbird, male, and apparently breeding or about to do so. I think this must be the north-west record point for this species.

Kispiox Valley, BC; 22 June 1921

The Audubon Warbler entered above is of interest, for although taken here at almost the extreme northern limit of the species, it is in some respects the most highly plumaged one I have ever taken. The black breast is nearly as lustrous and extensive as in *nigriferous*. At each corner of the yellow throat, tho, there is a distinct indication of a white mark which I do not recall as present in more southern specimens, and the wing is crossed by two white bars, as in *coronata*, not by a solid white patch. It will be of interest to find any indication of hybridization between the two species at this point, where they meet.

    . . . The Juncos are another group to be watched for anything of that sort. Today I shot a male of the *hyemalis* persuasion, probably the same as the Stikine River *J. h. connecteus*. There was a pair of them, making such a fuss, and acting otherwise in such manner as to convince me that they have a nest

Hazelton, British Columbia, 1921. Photograph by Harry S. Swarth, with permission of the Museum of Vertebrate Zoology, University of California, Berkeley.

somewhere nearby. Their calls attracted a pair of the common *Junco oreganus* (*shufeldti*), who were in the same bush, where I could compare them at my leisure. I must look for juncos and for junco's nests.

Kispiox Valley, BC; 8 July 1921

The Chestnut-collared Longspur listed above was shot by Strong in the road near here. This seems to me rather a remarkable occurrence. It looks like a breeding bird, but we have no others, nor have I heard any unfamiliar song such as that of the male bird must be. There are broad meadows here that might afford breeding sites, but no prairie land such as I thought this species always frequented.

[*Harry Swarth and William D. Strong stayed in a dilapidated cabin on Nine Mile Mountain while searching for ptarmigan and grouse on the ridges above the Skeena River. For two weeks they had persistent, unwanted cabin mates. —C. Swarth*]

Nine Mile Mountain, BC; 23 July 1921

At about 6:00 p.m. we were within half a mile of our destination but we lost the trail and that last half mile cost us about two hours. It was raining quite hard, and we had a long, steep hillside to climb. Horses and men were all in a state of exhaustion when we reached the cabin. And what a place! Dark, tumbledown and filthy beyond description. The house filled up with trash accumulated by

Rocher Déboulé Range, British Columbia, 1921. Photograph by Harry S. Swarth, with permission of the Museum of Vertebrate Zoology, University of California, Berkeley.

Nine Mile Mountain, British Columbia, 1921. Photograph by Harry S. Swarth, with permission of the Museum of Vertebrate Zoology, University of California, Berkeley.

the bushy-tailed woodrats. Porcupines had helped in the mess, and wood work was chewed up and quills and droppings scattered everywhere. About 11:00 p.m. we got to bed. As soon as we lay down and became quiet the woodrats became busy. They were quite fearless, and extraordinarily noisy. Once Frank (our packer) was waked by one scrambling over his head. The rats had chewed Frank's shoe strings into bits, and had eaten the seat out of his pants! We were dead tired and did not wake up until 8:00 o'clock.

21 July 1921

Another lovely night. About three o'clock I was awakened by a worse clamor than usual, and pushed aside the mosquito tent to see in the dim morning light a huge porcupine going up our shelves as tho ascending a fire escape. There are tiers of boxes nailed on the wall in which we keep provisions, and he went clear to the top, ascending a cascade of tinned goods. Milk, corned beef, etc., came down with a prodigious clatter, but he went up as tho climbing a rock slide, mounted the highest box, clear against the roof, and turned his back to us, daring us to do our worst. All attempts to dislodge him proving futile, I finally shot him with two aux, behind the shoulder. This did not at once kill him, but he came down, wedged in the corner behind the boxes, and with his tail projecting from the angle. There we had to leave him until morning, when with some difficulty he was dislodged.

24 July 1921

It would be most delightful to have a peaceful night's sleep once more. Yesterday we accounted for four woodrats and last evening one more before going to bed, but there are still some noisy survivors. We took the precaution of fastening the door before going to bed and it was well we did so, for another porcupine appeared. He wandered about the outside of the cabin endeavoring to gnaw an entry at various points. The noise he made was damnable. Finally, he crawled up to the partly blocked window and went to work, and as he was outlined against the sky Strong shot at him with his revolver. He fell with a thud, and Strong going out to examine the results found him apparently dead or dying, but this morning he was gone.

25 July 1921

Another blooming porcupine! We had the door closed, but he came around in the middle of the night and made such an infernal racket outside that Strong finally got up and shot him. Yesterday afternoon I saw one prowling about the cabin on the other side of the gulch, and in the evening Strong saw another in the rock slide behind us. Apparently, there is an endless supply, and apparently, they all have a mania for visiting our cabin.

26 July 1921

Another porcupine last night of course. I heard him moving about for some time, gnawing at all corners of the cabin, and finally when it became light

North American porcupine, 2011. Photograph by Lisa Hupp, US Fish and Wildlife Service.

Field assistant William
Duncan Strong, Kispiox
Valley, British Columbia,
1921.
Photograph by Harry S.
Swarth, with permission of
the Museum of Vertebrate
Zoology, University of
California, Berkeley.

enough to see I chased him away with a stick. He moved on down the trail, but reluctantly and with many backward glances.

27 July 1921
Naturally we had another porcupine visitor last night.

29 July 1921
This living in a haunted house is an infernal nuisance. Three porcupines visited the cabin last night and the noise they made was sinful.

5 August 1921
A visitation of porcupines last night that made all previous occasions of the kind tame and uninteresting. There were at least three, perhaps more, and none would leave until Strong went out with his six-shooter, and with bullets and naval vituperation shattered the calm of the night. He thinks he killed all of them, but we found just one below the cabin this morning. This was the last of the lot (he showed up just as dawn was breaking), and when Strong,

with a torch in his hand, opened the door he rushed right in. A whale of a big fellow, and he had the occupants of the house panic stricken, but Strong shot at him twice and he scrambled out and rolled down the hill, dead. This animal is extremely dark colored, while the other two I have handled have a great deal of yellow over hair. I have my doubts as to the existence of a coast race, at least on the basis of dark color. . . .

. . . Let me set down my opinion here that there is nothing more maddening than to be awakened from sound slumber by the monotonous—scrunch, scrunch, scrunch—of a porcupine gnawing at the corner of the house. It is an insistent, monotonous sound that forbids sleep, and there is no hope of his quitting and going home before daybreak. The porcupine is, without exception, the damnedest fool in the animal world.

6 August 1921

Only one porcupine visitor last night. He got his nose in a rat trap and left early.

[*Notes on a Merlin (Pigeon Hawk) Swarth collected*]:

Kispiox Valley, BC; 22 August 1921

Falco columbarius; Immature male. Feet bright yellow; claws black. Eye dark. Bare skin about eye greenish yellow. Cere greenish-yellow. Bill mostly blackish, bluish over basal half of lower mandible. Stomach contained a mass of bones and feathers—of Black Swift!

Kispiox Valley, BC; 23 August 1921

This morning a flock of Siskins flew past me with a Pigeon Hawk in pursuit. The speed this species attains is marvelous. This bird darted in and out among the trees, going so fast and with wings held so close to its body, that he did not really look much bigger than the birds he was after. As far as I could see, the only reason he failed to catch a Siskin was because he lacked the ability to concentrate! There were about 70 or 80 Siskins in all directions and he was trying to catch them all. He swept past me several times and I finally got him.

. . . Last evening several Horned Owls began hooting near camp just at dusk, and one lit at the top of a small spruce where we could see him against the sky. I had always imagined the hooting a rather dignified performance with the bird sitting sedately upright, until Strong described the antics of some he has watched lately. This bird hooted again and again, and each time the notes were pumped out with a series of jumping-jack like contortions. He leaned

Pencil sketch of a Great Horned Owl, by Allan Brooks, Atlin, British Columbia. Swarth family collection.

far forward, wings were raised and half spread, and the tail was tilted up at as absurd an angle as a Winter Wren could achieve.

Kispiox Valley, BC; 26 August 1921

The Rusty Blackbird is I think the most secretive of his kind. There are some around I know, but we very seldom see them. Yesterday afternoon I was watching the slough where I shot the muskrat, on the chance of seeing another, when I heard a slight splashing in the water. In a few moments a small bird appeared, walking in the reeds where it was nearly hidden. The movement of raising the gun startled it, and it gave a squawk and darted into thicker cover. It was all so distinctly rail-like that I half believed it must have been a smaller rail. A few minutes later the bird reappeared, a Rusty Blackbird, and went on feeding but a few yards from me. He was wading in water belly-deep, and darting after things on the surface. He worked his way upstream beyond my sight. If this is a usual mode of feeding it is no wonder that we see so few of the birds.

Hazelton, BC; 24 September 1921

On my way back I saw a Pigeon Hawk in a tree bordering one of Beirne's fields. Got the gun from the cabin and started back to where I saw the bird, but I had just entered the field when he appeared circling about. In a few moments he swung over in my direction and passed directly overhead but very high up, fifty yards I should say. I risked a load of 5's and luckily dropped him, the first in adult plumage secured on this trip—or on any other Museum expedition either, for that matter.

1924 Field Journal Entries

Carcross, Yukon Territory; 24 May 1924

Suddenly, toward the west, I saw a large flock of ducks coming from high over the mountains, and as they passed overhead could see they were White-winged Scoters, on migration inland. There were 60 or 70, a few uttering harsh "quacks"; they circled about once, then went on. It is one thing to read of their migration from the ocean, and quite another to actually see them. This was my first sight of this interesting flight.

Atlin, BC; 3 June 1924

Spent half an hour or more watching waxwings at a point just south of Pine Creek, where we always see them. Was rewarded by finding one nest, not finished but with the walls partly built, so that I could just see through. It was in a Jack Pine (*P. contorta*) about thirty feet up, and about three feet from the trunk. Then at another place in an open grove of smaller pines I watched three or four pair of waxwings. They were fussing about for all the world like young couples investigating apartment houses, examining tree after tree, clambering about among the middle branches close to the trunk, and buzzing sociably to

Major Allan Brooks, who assisted Swarth in the field in Atlin, BC. Swarth family collection.

Lake Atlin, 1924. Photograph by Harry S. Swarth, with permission of the Museum of Vertebrate Zoology, University of California, Berkeley.

each other at all times. One came with a mouthful of white fiber and tucked it into a crotch almost over my head. I don't believe she was seriously starting a nest, but four or five others (her sisters, and her cousins, and her aunts, apparently) sat around most interestedly watching how she did it. They are assuredly most delightful birds—gentle, companionable, and beautiful, and with so many interesting ways.

Atlin, BC; 15 June 1924

The Short-billed Gull was shot by myself, and Brooks shot another of that species and one Bonaparte at the same place. These all appeared to be non-breeding birds. The two Short-bills had been feeding on flying ants, a species abundant here now, black and about half an inch long. Each one had gullet and stomach filled to distension and each one had dead ants clinging to the inside of the mouth and throat, adhering with clamped jaws. A most uncomfortable meal, I should say.

Atlin, BC; 19 June 1924

A day devoted to Golden-crowned Sparrows, largely on the mountain we have visited before, a few miles south of Atlin. The summit is about 4,500 feet. . . . We knew the birds were there, for we had seen them before, but they were not much in evidence today. Half an hour search brought no results. . . . Then,

Atlin main street with islands in background, 1929. Photograph by Winifern Swarth.

as I was traversing a dry ridge, a sparrow darted out as I passed a low thicket, and a very little search revealed the nest, with five eggs! (This is perhaps the first authentic nest of this species to be discovered. I must look up the matter later.) The thicket the nest was in was a mixture of prostrate balsam and dwarf birch, very low, the branches rising not more than six or eight inches from the ground, and the cover rather sparse. There was a low ledge of protruding rock covered by this growth, and the nest was on the ground in the shelter of the rock and fairly well concealed by the overhanging vegetation.

Islands in Lake Atlin (near town), BC; 15 July 1924

A second notable feature of the islands is the abundance of birds, this in striking contrast to conditions on the mainland. Brooks listed 15 species apparently nesting on the third island. To our surprise the Blackpoll Warbler was about the most abundant small bird. Brooks's theory is (and it seems a self-evident proposition) that the abundance of small birds on the islands is due to the absence of squirrels, chipmunks, and Canada Jays, which persecute them on the mainland. Certainly, it is a striking fact that the island birds (except the waxwings) were all busily engaged in feeding young, whereas on the mainland any nest we have watched was almost sure to be destroyed by something.

Atlin, BC; 28 July 1924

Brooks's most noteworthy observation was of a Gyrfalcon. It was feeding upon a ground squirrel (*Citellus plesius*), and from "sign" about certain rock piles, had evidently occupied the region for some time. Brooks is satisfied that it is a breeding bird. It is ideal country for the species, with limitless expanses of open plateau (above timberline) extending north and east, and the valleys separating the ridges are relatively narrow and of no moment to a bird of such flying power.

Spruce Mountain, Atlin, BC; 3 August 1924

On the further slope of the valley, we found a fair number of Brewer's (?) Sparrows and immediately bent every effort and spent a good deal of time in their pursuit. They were in the scrubby trailing birch, a thick chaparral waist high in many places, and this brush had to be crashed through to flush the birds. They skulk deep in the vegetation, sitting shots are almost out of the question, and they are extremely hard to hit as they dart forth in zig-zag flight. In all, during the day we got six between us. With them were some Golden-crown and some Tree Sparrows.

Monarch Mountain, Atlin, BC; 5 September 1924

Halfway up the mountain a Goshawk came into view and when I "squeaked" he whirled about and came over me, when I shot him. This bird is in the adult plumage, and is very pale colored with fine vermiculations underneath, but it is noteworthy that it is just finishing the molt from the streaked, brown immature plumage. So it does not follow that the coarse, heavily marked, *striatus* type of bird represents a stage following the streaked juvenal, the finely marked, pale-colored birds, the fully mature. This argument has been advanced by Taverner and, I think, Bishop. On this particular specimen there are only a few juvenal feathers left, but they are there if searched for. It is an important point.

On Atlin Lake, BC; 7 September 1924

Shortly after we started 8 moose appeared swimming across the lake, but on getting nearer they took to their wings and flew away. White-winged Scoters they were, but in the misty light the deception was perfect.

Expedition Maps

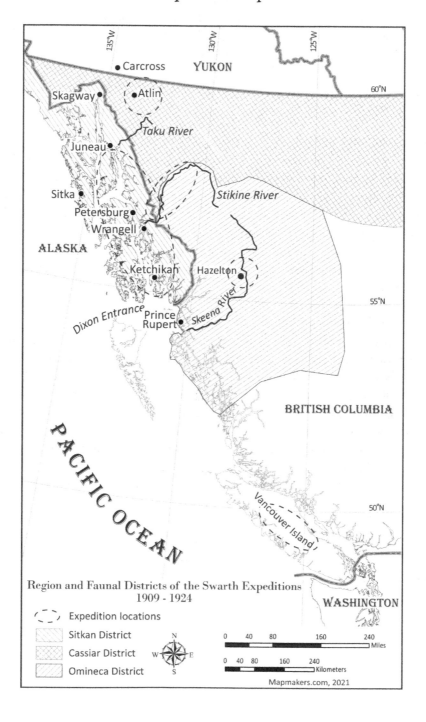

Region and Faunal Districts of the Swarth Expeditions
1909 - 1924

- ⌒‿⌒ Expedition locations
- ▨ Sitkan District
- ▨ Cassiar District
- ▨ Omineca District

Mapmakers.com, 2021

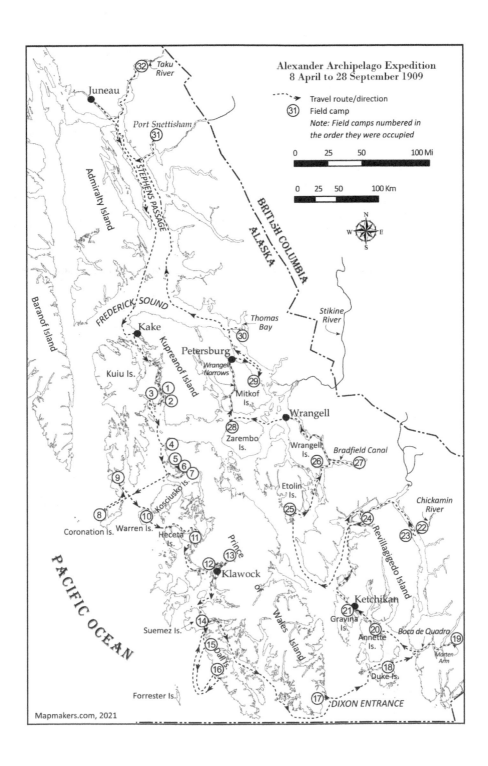

Alexander Archipelago Expedition
8 April to 28 September 1909

Travel route/direction
31 Field camp
Note: Field camps numbered in
the order they were occupied

0   25   50   100 Mi

0   25   50   100 Km

Mapmakers.com, 2021

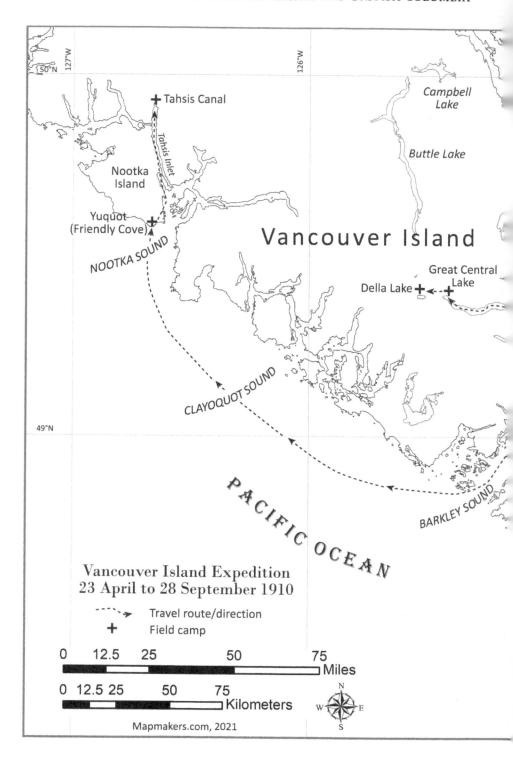

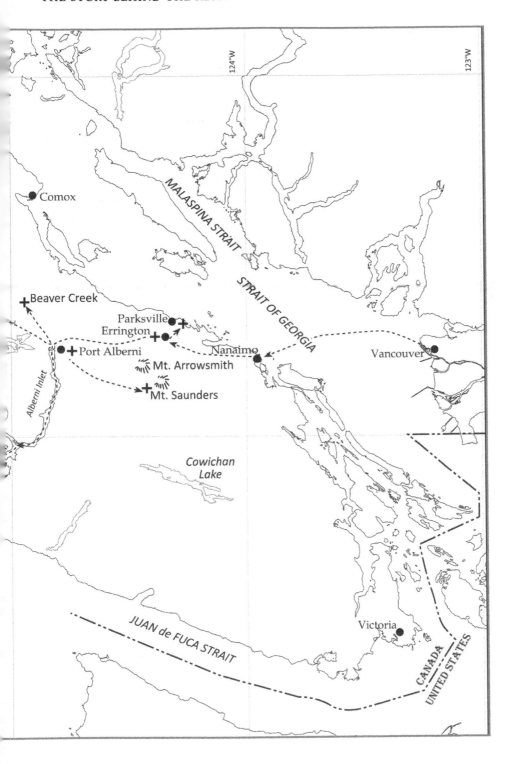

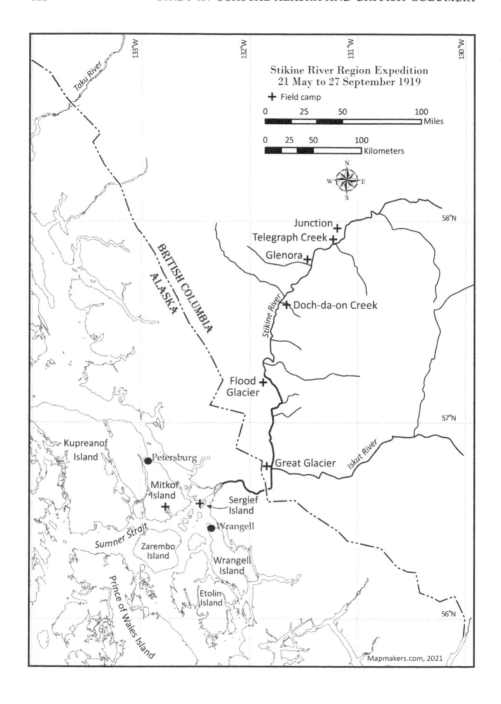

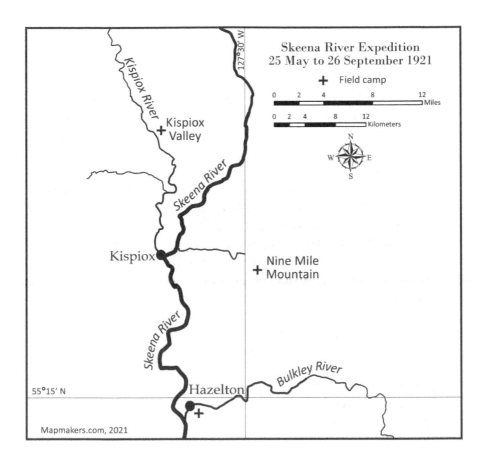

Expedition Itineraries

Alexander Archipelago and Southeastern Alaska Coast: April–September 1909

Juneau (8 April)
Kake, Kupreanof Island (9 April)
Keku Strait, Kupreanof Island (two camps, 10–18 and 18–25 April)
Three-Mile Arm, Kuiu Island (25 April–6 May)
Port Protection, Prince of Wales Island (6–10 May)
Shakan and Calder Bay, Prince of Wales Island (10–14 May)
Egg Harbor, Coronation Island (14–18 May)
Port McArthur, Kuiu Island (18–19 May)
Warren Island (19–23 May)
Heceta Island (23–24 May)
San Alberto Bay, Prince of Wales Island (24–26 May)
Klawak Salt Lake, Prince of Wales Island (26–29 May)
Suemez Island (29–30 May)
Bobs Bay, Dall Island (30 May–4 June)
Dall Island (4–6 June)
McLean Arm, Prince of Wales Island (6–7 June)
Duke Island (7–9 June)
Marten Arm, Boca de Quadra (9–14 June)
Annette Island (14–15 June)
Ketchikan and Gravina Island (15–16 June)
Chickamin River mouth (17–28 June)
Portage Cove, Revillagigedo Island (28 June–4 July)
Hassler Pass, Revillagigedo Island (4–5 July)
Etolin Island (5–12 July)
Wrangell Island (12–18 July)
Bradfield Canal (18–26 July)
Zarembo Island (27 July–1 August)
Mitkof Island (1–13 August)
Thomas Bay (13–23 August)
Port Snettisham (24 August–2 September)
Juneau (2–3 September)
Taku River mouth (4–28 September)
Juneau (28 September)

Vancouver Island: April–September 1910

Nanaimo (23–24 April)
Parksville (24 April–7 May)
Little Qualicum River (7–16 May)
French Creek (16–20 May)
Errington (20–27 May)
Alberni and Beaver Creek (27 May–28 June; Swarth arrived on 9 June)
Golden Eagle Mine (1–20 July)
Alberni (20–21 July)
Friendly Cove, Nootka Sound (22–23 July)
Tahsis Canal (24 July–2 August)
Friendly Cove, Nootka Sound (2–11 August)
Alberni (12–16 August)
Great Central Lake (17–19 August)
Della Lake (19–21 August)
Great Central Lake (21–26 August)
Alberni (26–29 August)
Errington (29 August–6 September)
Mount Arrowsmith (6–8 September)
Errington (8–28 September)

Stikine River Region: May–September 1919

Wrangell (21 May)
Telegraph (23 May)
The Junction (25 May–6 June)
Telegraph Creek (8–26 June)
Glenora (27 June–8 July)
Doch-da-on Creek (8–26 July)
Flood Glacier (26 July–8 August)
Great Glacier (8–16 August)
Sergief Island (17 August–7 September)
Mitkof Island (26–29 August; J. Dixon only)

Skeena River Region: May–September 1921

Hazelton (25 May–20 June)
Kispiox Valley (20 June–15 July)
Hazelton (15–21 July)
Nine Mile Mountain (21 July–14 August)
Kispiox Valley (16 August–17 September)
Hazelton (17–26 September)

Atlin Region: May–September 1924

Skagway (21 May)
Carcross (21–27 May)
Atlin (27 May–26 July)
Otter Creek (26 July–9 August)
Atlin (9 August–7 September)
Teslin Lake (7–14 September)
Atlin (15–24 September)

SECTION 3

Bird Species Checklists

Selected lists of birds that Harry Swarth observed on his expeditions to the Pacific Northwest are shown here. These checklists, using current common names, were compiled from Swarth's field journals, which are archived at the Museum of Vertebrate Zoology, and from his published reports. They illustrate the composition of the bird communities in areas where Swarth spent considerable time exploring and collecting.

Table 1. Kupreanof Island, Southeastern Alaska. 10–24 April 1909.

White-fronted Goose	Common Merganser	chickadee (unid.)
Canada Goose	grouse (unid.)	Brown Creeper
Mallard	Greater Yellowlegs	Golden-crowned Kinglet
Northern Pintail	Marbled Murrelet	Ruby-crowned Kinglet
Long-tailed Duck	loon (unid.)	American Robin
Bufflehead	Great Blue Heron	Varied Thrush
goldeneye (unid.)	Sharp-shinned Hawk	American Pipit
scaup (unid.)	Bald Eagle	Common Redpoll
Harlequin Duck	Red-breasted Sapsucker	White-winged Crossbill
Surf Scoter	Northern Flicker	Song Sparrow
White-winged Scoter	Steller's Jay	Dark-eyed ("Oregon") Junco
Black Scoter	Common Raven	
Long-tailed Duck	Northwestern Crow	

Table 2. Egg Harbor, Coronation Island, Southeastern Alaska. 14–18 May 1909.

Mallard	Savannah Sparrow	Northwestern Crow
Harlequin Duck	Pigeon Guillemot	Common Raven
White-winged Scoter	Marbled Murrelet	Chestnut-backed Chickadee
Long-tailed Duck	Rhinoceros Auklet	Red-breasted Nuthatch
Bufflehead	Glaucous-winged Gull	Pacific Wren
Common Merganser	loon (unid.)	Golden-crowned Kinglet
Rufous Hummingbird	cormorant (unid.)	Ruby-crowned Kinglet
Black Oystercatcher	Sharp-shinned Hawk	Hermit Thrush
Semipalmated Plover	Bald Eagle	American Robin
Least Sandpiper	Hairy Woodpecker	Varied Thrush
Western Sandpiper	Pacific-slope Flycatcher	
Wandering Tattler	Steller's Jay	

Table 3. Chickamin River, Southeastern Alaska. 17–28 June 1909.

Canada Goose	Alder Flycatcher	Cedar Waxwing
Common Merganser	Pacific-slope Flycatcher	Pine Grosbeak
Sooty Grouse	Steller's Jay	Red Crossbill
Vaux's Swift	Tree Swallow	Pine Siskin
Rufous Hummingbird	Violet-green Swallow	Savannah Sparrow
Spotted Sandpiper	Barn Swallow	Song Sparrow
Sharp-shinned Hawk	Chestnut-backed Chickadee	Lincoln's Sparrow
Bald Eagle	Pacific Wren	MacGillivray's Warbler
Red-tailed Hawk	Golden-crowned Kinglet	Common Yellowthroat
Belted Kingfisher	Ruby-crowned Kinglet	Yellow Warbler
Red-breasted Sapsucker	Swainson's ("Russet-backed") Thrush	Western Tanager
Hairy Woodpecker	American Robin	
Western Wood-Pewee	Varied Thrush	

Table 4. Portage Cove, Revillagigedo Island, Southeastern Alaska. 28 June– 4 July 1909.

Canada Goose	Hairy Woodpecker	American Robin
Mallard	Northern (*auratus* × *cafer*) Flicker	Varied Thrush
Harlequin Duck	Pacific-slope Flycatcher	Pine Grosbeak
Common Merganser	Steller's Jay	Red Crossbill
Black Swift	Northwestern Crow	Lincoln's Sparrow
Vaux's Swift	Tree Swallow	Song Sparrow
Spotted Sandpiper	Barn Swallow	Dark-eyed ("Oregon") Junco
Marbled Murrelet	Chestnut-backed Chickadee	Orange-crowned Warbler
Bonaparte's Gull	Pacific Wren	Wilson's Warbler
Sharp-shinned Hawk	Swainson's ("Russet-backed") Thrush	
Bald Eagle	Hermit Thrush	

Table 5. Mitkof Island, Southeastern Alaska. 1–13 August 1909.

Sooty Grouse	Great Horned Owl	Hermit Thrush
Spotted Sandpiper	Northern Saw-whet Owl	American Robin
Wandering Tattler	Belted Kingfisher	Varied Thrush
murrelet (unid.)	Pacific-slope Flycatcher	Song Sparrow
Bonaparte's Gull	Steller's Jay	Lincoln's Sparrow
Short-billed Gull	Tree Swallow	Dark-eyed ("Oregon") Junco
Glaucous-winged Gull	Chestnut-backed Chickadee	Orange-crowned Warbler
Great Blue Heron	Pacific Wren	Townsend's Warbler
Sharp-shinned Hawk	Golden-crowned Kinglet	Wilson's Warbler
Bald Eagle	Ruby-crowned Kinglet	

Table 6. Thomas Bay, Southeastern Alaska. 13–23 August 1909.

Canada Goose	Spotted Sandpiper	Violet-green Swallow
Mallard	Marbled Murrelet	Bank Swallow
Green-winged Teal	Bonaparte's Gull	Barn Swallow
scaup (unid.)	Short-billed Gull	Chestnut-backed Chickadee
Surf Scoter	Glaucous-winged Gull	Golden-crowned Kinglet
Black Scoter	Arctic Tern	Ruby-crowned Kinglet
White-winged Scoter	loon (unid.)	Hermit Thrush
Common Merganser	Great Blue Heron	American Robin
Horned Grebe	Sharp-shinned Hawk	Common Redpoll
Vaux's Swift	Northern Goshawk	Red Crossbill
Rufous Hummingbird	Bald Eagle	Savannah Sparrow
Sandhill Crane	Great Horned Owl	Song Sparrow
Semipalmated Plover	Belted Kingfisher	Lincoln's Sparrow
Baird's Sandpiper	Merlin	Golden-crowned Sparrow
Pectoral Sandpiper	Olive-sided Flycatcher	Dark-eyed ("Oregon") Junco
Least Sandpiper	Western Wood-Pewee	Orange-crowned Warbler
Western Sandpiper	Pacific-slope Flycatcher	Townsend's Warbler
Lesser Yellowlegs	Northwestern Crow	Yellow Warbler
Greater Yellowlegs	Common Raven	Wilson's Warbler
Wilson's Snipe	Tree Swallow	

Table 7. Taku River (20 miles inland from the mouth), Southeastern Alaska. 4–28 September 1909.

goose (unid.)	Long-eared Owl	American Pipit
Mallard	Short-eared Owl	Lapland Longspur
Green-winged Teal	Downy Woodpecker	Savannah Sparrow
Willow Ptarmigan	Northern Flicker	Fox Sparrow
Black-bellied Plover	Merlin	Song Sparrow
Pectoral Sandpiper	Steller's Jay	Lincoln's Sparrow
Wilson's Snipe	Black-billed Magpie	White-crowned Sparrow
Spotted Sandpiper	Northwestern Crow	Golden-crowned Sparrow
Solitary Sandpiper	Common Raven	Dark-eyed ("Slate-colored") Junco
Bonaparte's Gull	Red-breasted Nuthatch	Dark-eyed ("Oregon") Junco
Glaucous-winged Gull	Pacific Wren	Orange-crowned Warbler
Herring Gull	Golden-crowned Kinglet	Common Yellowthroat
Osprey	Ruby-crowned Kinglet	Blackpoll Warbler
Northern Harrier	Mountain Bluebird	Townsend's Warbler
Sharp-shinned Hawk	Hermit Thrush	Yellow Warbler
Northern Goshawk	American Robin	Yellow-rumped ("Myrtle") Warbler
Bald Eagle	Varied Thrush	Wilson's Warbler

Table 8. Beaver Creek, Alberni, Vancouver Island, British Columbia. 9–29 June 1910.

Common Merganser	Pileated Woodpecker	Townsend's Solitaire
Ruffed Grouse	American Kestrel	Swainson's ("Russet-backed") Thrush
Sooty Grouse	Olive-sided Flycatcher	American Robin
Band-tailed Pigeon	Western Wood-Pewee	Varied Thrush
Mourning Dove	Alder Flycatcher	Cedar Waxwing
Common Nighthawk	Hammond's Flycatcher	Red Crossbill
Vaux's Swift	Warbling Vireo	Pine Siskin
Rufous Hummingbird	Red-eyed Vireo	Spotted Towhee
Virginia Rail	Steller's Jay	Song Sparrow
Turkey Vulture	Common Raven	Dark-eyed ("Oregon") Junco
Bald Eagle	Violet-green Swallow	Western Meadowlark
Red-tailed Hawk	Chestnut-backed Chickadee	Brewer's Blackbird
Great Horned Owl	Red-breasted Nuthatch	Common Yellowthroat
Belted Kingfisher	Brown Creeper	Yellow Warbler
Red-breasted Sapsucker	House Wren	Western Tanager
Hairy Woodpecker	Pacific Wren	Black-headed Grosbeak
Northern Flicker	Golden-crowned Kinglet	

Table 9. Nootka Sound, Vancouver Island, British Columbia. 22 July–11 August 1910.

Harlequin Duck	Great Blue Heron	Pacific Wren
White-winged Scoter	Osprey	American Dipper
goldeneye (unid.)	Sharp-shinned Hawk	Golden-crowned Kinglet
Common Merganser	Bald Eagle	Swainson's ("Russet-backed") Thrush
Ruffed Grouse	Belted Kingfisher	Hermit Thrush
Band-tailed Pigeon	Hairy Woodpecker	American Robin
Black Swift	Northern Flicker	Varied Thrush
Rufous Hummingbird	Olive-sided Flycatcher	Cedar Waxwing
Semipalmated Plover	Alder Flycatcher	Fox Sparrow
Least Sandpiper	Pacific-slope Flycatcher	Song Sparrow
Western Sandpiper	Steller's Jay	Orange-crowned Warbler
Spotted Sandpiper	Northwestern Crow	MacGillivray's Warbler
Common Murre	Common Raven	Yellow Warbler
Pigeon Guillemot	Violet-green Swallow	Townsend's Warbler
Marbled Murrelet	Chestnut-backed Chickadee	Wilson's Warbler
Glaucous-winged Gull	Red-breasted Nuthatch	
Common Loon	Brown Creeper	

Table 10. Great Central Lake, Vancouver Island, British Columbia. 17–26 August 1910.

Mallard	Osprey	Chestnut-backed Chickadee
Bufflehead	Northern Goshawk	Pacific Wren
White-tailed Ptarmigan	Northern Pygmy-Owl	American Dipper
Pied-billed Grebe	Merlin	Pine Siskin
Rufous Hummingbird	Hairy Woodpecker	Dark-eyed Junco
Bonaparte's Gull	Hammond's Flycatcher	
loon (unid.)	Canada Jay	

Table 11. Telegraph Creek, British Columbia. 8–25 June 1919.

Mallard	Northern Flicker	Pine Siskin
Green-winged Teal	Olive-sided Flycatcher	Chipping Sparrow
scaup (unid.)	Western Wood-Pewee	Song Sparrow
Harlequin Duck	Alder Flycatcher	Lincoln's Sparrow
White-winged Scoter	Dusky Flycatcher	White-crowned Sparrow
Barrow's Goldeneye	Say's Phoebe	Dark-eyed ("Oregon") Junco
Ruffed Grouse	Common Raven	Rusty Blackbird
Horned Grebe	Warbling Vireo	Tennessee Warbler
Mourning Dove	Violet-green Swallow	Orange-crowned Warbler
Common Nighthawk	Black-capped Chickadee	MacGillivray's Warbler
Black Swift	Mountain Bluebird	American Redstart
Spotted Sandpiper	Townsend's Solitaire	Yellow Warbler
Lesser Yellowlegs	Swainson's ("Olive-backed") Thrush	Yellow-rumped ("Myrtle") Warbler
Common Loon	American Robin	Wilson's Warbler
Red-tailed Hawk	Bohemian Waxwing	Western Tanager
Red-breasted Sapsucker	Pine Grosbeak	
Hairy Woodpecker	Red Crossbill	

Table 12. Doch-da-on Creek, British Columbia. 8–26 July 1919.

Ruffed Grouse	Northern ("Yellow-shafted") Flicker	Bohemian Waxwing
Dusky Grouse	American Kestrel	American Pipit
White-tailed Ptarmigan	Alder Flycatcher	Pine Grosbeak
Common Nighthawk	Hammond's Flycatcher	Gray-crowned ("Hepburn's") Rosy-Finch
Vaux's Swift	Warbling Vireo	White-winged Crossbill
Rufous Hummingbird	Steller's Jay	Pine Siskin
Short-billed Gull	Horned Lark	Chipping Sparrow
Arctic Tern	Black-capped Chickadee	Fox Sparrow
Common Loon	Mountain Chickadee	Song Sparrow
Northern Goshawk	Red-breasted Nuthatch	Lincoln's Sparrow
Bald Eagle	Golden-crowned Kinglet	Golden-crowned Sparrow
Red-tailed Hawk	Ruby-crowned Kinglet	Dark-eyed ("Slate-colored") Junco
Great Horned Owl	Mountain Bluebird	Rusty Blackbird
Northern Pygmy-Owl	Townsend's Solitaire	MacGillivray's Warbler
Belted Kingfisher	Swainson's ("Olive-backed") Thrush	American Redstart
Yellow-bellied Sapsucker	Hermit Thrush	Yellow Warbler
Red-breasted Sapsucker	American Robin	Yellow-rumped ("Myrtle") Warbler
Hairy Woodpecker	Varied Thrush	

Table 13. Sergief Island, Alaska. 17 August–5 September 1919.

Canada Goose	Short-eared Owl	Varied Thrush
Mallard	American Kestrel	American Pipit
Northern Pintail	Merlin	Red Crossbill
Green-winged Teal	Peregrine	White-winged Crossbill
Northern Shoveler	Pacific-slope Flycatcher	Lapland Longspur
Mourning Dove	Say's Phoebe	Savannah Sparrow
Black Swift	Steller's Jay	Fox Sparrow
Rufous Hummingbird	Northwestern Crow	Song Sparrow
Least Sandpiper	Common Raven	Lincoln's Sparrow
Pectoral Sandpiper	Horned Lark	Golden-crowned Sparrow
Semipalmated Sandpiper	Violet-green Swallow	Dark-eyed ("Oregon") Junco
Western Sandpiper	Bank Swallow	Rusty Blackbird
Wilson's Snipe	Barn Swallow	Orange-crowned Warbler
Greater Yellowlegs	Brown Creeper	MacGillivray's Warbler
jaeger (unid.)	Pacific Wren	Yellow Warbler
Great Blue Heron	Ruby-crowned Kinglet	Yellow-rumped ("Myrtle") Warbler
Northern Harrier	Swainson's ("Russet-backed") Thrush	Wilson's Warbler
Sharp-shinned Hawk	Hermit Thrush	
Steller's Sea Eagle	American Robin	

Table 14. Hazelton, British Columbia. 25 May–20 June 1921.

Ruffed Grouse	Western Wood-Pewee	American Pipit
Black Swift	Alder Flycatcher	Evening Grosbeak
Vaux's Swift	Hammond's Flycatcher	Purple Finch
Rufous Hummingbird	Warbling Vireo	Pine Siskin
Great Blue Heron	Red-eyed Vireo	Chipping Sparrow
Golden Eagle	American Crow	Song Sparrow
Sharp-shinned Hawk	Tree Swallow	White-crowned Sparrow
Northern Goshawk	Violet-green Swallow	Dark-eyed ("Slate-colored") Junco
Great Horned Owl	Northern Rough-winged Swallow	Dark-eyed ("Oregon") Junco
Belted Kingfisher	Cliff Swallow	Rusty Blackbird
Red-breasted Sapsucker	Black-capped Chickadee	Orange-crowned Warbler
American Three-toed Woodpecker	Red-breasted Nuthatch	MacGillivray's Warbler
Hairy Woodpecker	Golden-crowned Kinglet	Magnolia Warbler
Northern ("Yellow-shafted") Flicker	Ruby-crowned Kinglet	American Redstart
Pileated Woodpecker	Mountain Bluebird	Yellow Warbler
American Kestrel	Hermit Thrush	Yellow-rumped ("Audubon's") Warbler
Merlin	American Robin	Wilson's Warbler
Olive-sided Flycatcher	Gray Catbird	Western Tanager

Table 15. Kispiox Valley, British Columbia. 20 June–15 July 1921.

Canada Goose	Western Wood-Pewee	White-winged Crossbill
Common Merganser	Alder Flycatcher	Pine Siskin
Ruffed Grouse	Dusky Flycatcher	Chestnut-collared Longspur
Common Nighthawk	Hammond's Flycatcher	Chipping Sparrow
Black Swift	Warbling Vireo	Savannah Sparrow
Vaux's Swift	Red-eyed Vireo	Song Sparrow
Rufous Hummingbird	Steller's Jay	Lincoln's Sparrow
Common Loon	Canada Jay	White-throated Sparrow
Spotted Sandpiper	Tree Swallow	White-crowned Sparrow
Great Blue Heron	Northern Rough-winged Swallow	Dark-eyed ("Slate-colored") Junco
Northern Harrier	Barn Swallow	Dark-eyed ("Oregon") Junco
Sharp-shinned Hawk	Black-capped Chickadee	Rusty Blackbird
Bald Eagle	Red-breasted Nuthatch	Northern Waterthrush
Red-tailed Hawk	House Wren	Tennessee Warbler
Great Horned Owl	Ruby-crowned Kinglet	MacGillivray's Warbler
Belted Kingfisher	Mountain Bluebird	American Redstart
American Three-toed Woodpecker	Townsend's Solitaire	Common Yellowthroat
Red-breasted Sapsucker	Swainson's ("Olive-backed") Thrush	Magnolia Warbler
Downy Woodpecker	American Robin	Yellow Warbler
Hairy Woodpecker	Varied Thrush	Yellow-rumped ("Audubon's") Warbler
Northern ("Yellow-shafted") Flicker	Bohemian Waxwing	Western Tanager
American Kestrel	Cedar Waxwing	
Eastern Kingbird	Purple Finch	
Olive-sided Flycatcher	Pine Grosbeak	

Table 16. Nine Mile Mountain, British Columbia. 21 July–14 August 1921.

White-tailed Ptarmigan	Boreal Chickadee	Gray-crowned ("Hepburn's") Rosy-Finch
Rock Ptarmigan	Mountain Chickadee	White-winged Crossbill
Spruce Grouse	Red-breasted Nuthatch	Pine Siskin
Blue Grouse	Brown Creeper	American Tree Sparrow
Rufous Hummingbird	Pacific Wren	Savannah Sparrow
Baird's Sandpiper	Golden-crowned Kinglet	Fox Sparrow
Short-billed Gull	Ruby-crowned Kinglet	Golden-crowned Sparrow
Sharp-shinned Hawk	Mountain Bluebird	Dark-eyed ("Oregon") Junco
Golden Eagle	Hermit Thrush	Orange-crowned Warbler
American Kestrel	Varied Thrush	Townsend's Warbler
Canada Jay	American Pipit	Wilson's Warbler
Horned Lark	Pine Grosbeak	

Table 17. Carcross, Yukon Territory. 21–27 May 1924.

Mallard	Lesser Yellowlegs	Barn Swallow
Northern Pintail	Red-necked Phalarope	Black-capped Chickadee
Green-winged Teal	Bonaparte's Gull	Boreal Chickadee
Barrow's Goldeneye	Short-billed Gull	Ruby-crowned Kinglet
Bufflehead	Herring Gull	Mountain Bluebird
Greater Scaup	Arctic Tern	Hermit Thrush
White-winged Scoter	Pacific Loon	American Robin
Common Merganser	Golden Eagle	Bohemian Waxwing
Spruce Grouse	Swainson's Hawk	Lapland Longspur
Horned Grebe	Northern ("Yellow-shafted") Flicker	Chipping Sparrow
Semipalmated Plover	Peregrine Falcon	White-crowned Sparrow
Surfbird	Olive-sided Flycatcher	Dark-eyed ("Slate-colored") Junco
Baird's Sandpiper	Western Wood-Pewee	Rusty Blackbird
Least Sandpiper	Say's Phoebe	Orange-crowned Warbler
Wilson's Snipe	Canada Jay	Yellow-rumped ("Myrtle") Warbler
Spotted Sandpiper	Black-billed Magpie	Wilson's Warbler
Solitary Sandpiper	Violet-green Swallow	
Wandering Tattler	Cliff Swallow	

Table 18. Atlin, British Columbia. 3 June 1924.

Green-winged Teal	Tree Sparrow	White-winged Crossbill
Lesser Yellowlegs	Violet-green Swallow	Chipping Sparrow
Northern ("Yellow-shafted") Flicker	Cliff Swallow	Savannah Sparrow
American Kestrel	Barn Swallow	Dark-eyed ("Slate-colored") Junco
Olive-sided Flycatcher	Boreal Chickadee	Yellow-rumped ("Myrtle") Warbler
Western Wood-Pewee	American Robin	

Table 19. Atlin, British Columbia. 13–22 August 1924.

Northern Pintail	American Three-toed Woodpecker	Swainson's ("Olive-backed") Thrush
scaup (unid.)	Hairy Woodpecker	American Robin
White-winged Scoter	Northern Flicker	White-winged Crossbill
Barrow's Goldeneye	American Kestrel	Pine Siskin
Horned Grebe	Merlin	Chipping Sparrow
Red-necked Grebe	Olive-sided Flycatcher	Savannah Sparrow
Common Nighthawk	Western Wood-Pewee	Lincoln's Sparrow
Semipalmated Plover	Hammond's Flycatcher	White-crowned Sparrow
Spotted Sandpiper	Dusky Flycatcher	Dark-eyed ("Slate-colored") Junco
Lesser Yellowlegs	Canada Jay	Northern Waterthrush
Common Loon	Violet-green Swallow	Orange-crowned Warbler
Pacific Loon	Cliff Swallow	Yellow Warbler
Sharp-shinned Hawk	Barn Swallow	Blackpoll Warbler
Northern Goshawk	Black-capped Chickadee	Yellow-rumped ("Myrtle") Warbler
Red-tailed Hawk	Boreal Chickadee	Townsend's Warbler
Rough-legged Hawk	Red-breasted Nuthatch	Wilson's Warbler
Northern Hawk Owl	Ruby-crowned Kinglet	
Great Horned Owl	Townsend's Solitaire	

Table 20. Atlin, British Columbia. 19 September 1924.

Mallard	Red-tailed Hawk	Mountain Bluebird
scaup (unid.)	Northern Hawk Owl	Hermit Thrush
White-winged Scoter	Canada Jay	White-winged Crossbill
Barrow's Goldeneye	Black-billed Magpie	Pine Siskin
Ruffed Grouse	Boreal Chickadee	Dark-eyed ("Slate-colored") Junco
Horned Grebe	Red-breasted Nuthatch	Yellow-rumped ("Myrtle") Warbler
Northern Goshawk	Ruby-crowned Kinglet	

SECTION 4

Ornithological Perspectives on Swarth's Research

A Modern Perspective on Landbird Population Expansions along the Coast

BY STEVE HEINL, KETCHIKAN

Swarth's thesis was that the avifauna of southeastern Alaska, which comprises a narrow area of mainland and the islands of the Alexander Archipelago, was blocked from the continental interior by the rugged Coast Range, which "stood as an impassable barrier" to immigration from the east following the retreat of glacial ice at the close of the Pleistocene (see also Swarth's 1936 paper on the origins of birds of the Sitkan district). The ranges of many continental birds extend west to the Coast Range but no farther. Swarth showed that species that occur in both southeastern Alaska and interior British Columbia are typically represented by different subspecies on each side of the Coast Range, and that the interior subspecies often range widely across the continent. Conversely, the avifauna of southeastern Alaska is largely coastal, consisting of species and subspecies restricted to rain-forest habitats throughout coastal southeastern Alaska and British Columbia as well as some near the northern extent of a coastal distribution extending north from coastal California. Thus, the avifauna of southeastern Alaska is primarily of southern origin, the only direction from which birds could move as a result of the regional geography. These faunal origins were made clear to Swarth as a result of his experience traveling in both regions and through his careful study of specimens (both bird and mammal), many of which he collected himself.

Knowledge of Alaska's avifauna has steadily advanced in the nearly ninety years since Swarth wrote this manuscript, and a host of species are known to have extended their ranges west through the Coast Range into mainland river systems in southeastern Alaska. These mainland river valleys support deciduous (cottonwood, willow, alder) and mixed coniferous/deciduous

Stikine River mouth, southeastern Alaska. Photograph by Sam Beebe, Ecotrust.

(cottonwood and spruce) riparian forests, interior-like habitats that are not found elsewhere in southeastern Alaska. Some species that Swarth outlined as birds of the interior are now common or fairly common breeders in mainland southeastern Alaska river valleys (e.g., Hammond's Flycatcher, Warbling Vireo, MacGillivray's Warbler, American Redstart, Yellow Warbler); others are uncommon (e.g., Ruffed Grouse, Rusty Blackbird, Northern Waterthrush, Western Tanager, Chipping Sparrow), while others are of rare or irregular occurrence (e.g., Least Flycatcher, Tennessee Warbler, Magnolia Warbler) (see Kessel and Gibson 1978; Johnson et al. 2008).

It is difficult to say whether some of these species were already present in Alaska in Swarth's time or whether some have since extended their ranges, although I suspect a combination of both. Swarth and other ornithologists who visited southeastern Alaska in the early twentieth century did not spend much time in mainland southeastern Alaska river valleys during the breeding season (Grinnell 1898, 1909; Willett 1915, 1923, 1927, 1928; Swarth 1911, 1922; Bailey 1927; and see Gabrielson and Lincoln 1959). Indeed, Swarth thought these connecting river systems provided "migration paths or rather by-paths, to a limited extent . . . used mostly by inexperienced young birds on their first southward flight." Still, it seems inevitable that species distributions would change. Several species that Swarth discussed as birds of the interior are now distributed in southeastern Alaska well beyond the mainland river systems (e.g., Yellow-rumped and Wilson's Warblers, Common Yellowthroat), and it seems unlikely that Swarth and other ornithologists would have over-looked them. I have personally observed range expansions of several species (e.g., Anna's Hummingbird, Barred Owl, Warbling Vireo, and White-crowned Sparrow subspecies *pugetensis*) in the thirty years that I've lived in southeast-ern Alaska. Of course, whether some species were overlooked in Swarth's time or whether, instead, range expansions have occurred in the intervening years does not detract at all from Swarth's thesis regarding the effect of glaciation and the Coast Range on the origins of southeastern Alaska birds.

I was fascinated to read some of Swarth's comments on the paucity of breeding birds and migrants in southeastern Alaska. He wrote, "The woods are generally hushed and still, seldom disturbed by song of bird or chirp of insect," and "the characteristic and penetrating call note of the Western Flycatcher is to be heard in the woods of even the westernmost islands of the Alexander Archipelago, where it may be almost the only bird sound to break the stillness." Swarth apparently met with only small numbers of migrant passerines during his work in the Alexander Archipelago in April and May 1909 and concluded that the islands of the archipelago "do not lie in the main migration route of the birds of the Pacific coast" (Swarth 1911:159). Grinnell (1898:122) similarly observed low abundance and diversity of birds in the coniferous forest of the Sitka area and concluded that "the dark mossy forests but a few rods back from the coast are almost destitute of bird life."

Warbling Vireo, California. Photograph by Mark A. Chappell.

I have a very different impression regarding abundance of migrant and breeding songbirds in Ketchikan, Alaska. During migration, American Robins, Varied Thrushes, Orange-crowned and Yellow Warblers, Ruby-crowned Kinglets, Golden-crowned, Fox, Lincoln's, and Savannah Sparrows, and others can at times be present in droves during April and early May, particularly during periods of foul weather (Heinl and Piston 2009); however, this abundance may very well be a result of the concentrating effect of altered habitats. Over the past ninety years, humans have logged forests, built roads, and developed towns. These alterations have provided new habitat in the form of deciduous trees (primarily alder), brushy areas, and vacant lots and open grassy areas at schools, parks, and airports, all of which provide oases for migrants (as well as breeding habitat) in the surrounding expanse of coniferous forest and muskeg.

I was recently reminded of my skewed view of the abundance of migrants when I spent a week in late April at Hugh Smith Lake, in the Misty Fjords National Monument Wilderness, where the habitat is essentially the untouched coniferous habitat that Swarth encountered during his travels in southeastern Alaska in 1909. Migrant songbirds were very common in Ketchikan the day I departed for my trip, but during my week at Hugh Smith Lake I saw only a handful of migrant warblers and sparrows in the surrounding coniferous forest. The forest, however, was not so dramatically "hushed" or "destitute of bird life" as Swarth and Grinnell described, as breeding species such as Rufous Hummingbird, Chestnut-backed Chickadee, Pacific Wren, Hermit Thrush, Varied Thrush, Townsend's Warbler, Song Sparrow, and others were conspicuous.

Leapfrog Migration in Fox Sparrows of the Pacific Northwest

PHILIP UNITT, SAN DIEGO NATURAL HISTORY MUSEUM

The wonders of bird migration conceal layer upon layer of complexity. We have Harry Swarth to thank for uncovering one of these layers and introducing the concept of "leapfrog migration." In essence, leapfrog migration is a pattern of migration in which the populations of a species breeding farthest north spend the winter the farthest south, leaping over populations that breed farther south but winter farther north. Swarth (1920) developed this concept through his study of geographic variation in the Fox Sparrow (*Passerella iliaca*), on both its breeding and winter ranges. He found that the brown-backed populations breeding in southwestern and south-central Alaska, the subspecies *P. iliaca unalaschcensis, insularis,* and *sinuosa,* are the dominant brown or "sooty" Fox Sparrows wintering in southern California. The subspecies *annectens,* breeding in the Yakutat area of Alaska, in winter dominates in the San Francisco Bay area. Subspecies *townsendi,* breeding in southeastern Alaska, in winter dominates in coastal Oregon, while the southernmost subspecies, *fuliginosa,* breeding on Vancouver Island and the Olympic Peninsula, is rarely collected far south of its breeding range.

Since the publication of Swarth's landmark study in 1920, other examples of leapfrog migration have been found in a wide diversity of birds, including the Bar-tailed Godwit (Drent and Piersma 1989), Rock Sandpiper (Boland 1990), and Wilson's Warbler (Kelly et al. 2002). Leapfrog migration is generally acknowledged to be a pattern more widespread than "chain migration," in which populations within a species maintain their same order from north to south in both summer and winter. More broadly, Swarth's example with the Fox Sparrow was an early step leading to the concept that currently goes by the rather opaque term "migratory connectivity." That is, species of migratory birds represent a spectrum, from birds that breed at one locality and disperse widely throughout the species' winter range (low connectivity), to birds that all breed at one locality and spend the winter within one small subset of the winter range (strong connectivity). Fidelity to a site within the winter range can be just as strong as fidelity to a site within the breeding range. It is the linking of these fidelities that gave rise to the term "connectivity."

Various explanations for the evolution of leapfrog migration have been put forward, including intraspecific competition, latitudinal variation in the predictability of the breeding site becoming suitable, and, finally, the balance between the cost and timing of migration and the availability of food along the

Fox Sparrow, Anchorage, Alaska.
Photograph by Mark A. Chappell.

route (Bell 1997). As the climate changes, this balance will also change. Fontaine et al. (2015) concluded that climate change will affect leapfrog migrants more strongly than chain migrants.

In spite of many subsequent studies of leapfrog migration, Harry Swarth's Fox Sparrows remain the classic example, and his figure H (page 135) has been reproduced repeatedly in ornithology textbooks and reference works on bird migration, as by Welty and Baptista (1988), Elphick (2007), Newton (2007), and Weckstein et al. (2020). One subsequent study, based on the recapture of Fox Sparrows that had been equipped with light-level geolocators, failed to corroborate the pattern of leapfrog migration among the Pacific coast birds (Fraser et al. 2018). It was based, however, on the capture of birds from only two sites and could have included migrants as well as birds wintering at the site of capture. Also, it is quite possible that the winter ranges of the various subspecies have shifted in the century since the specimens on which Swarth based his study were collected, thus muddying the pattern. As far as southern California is concerned, however, the basic pattern Swarth outlined persists today: the northernmost subspecies winter commonly, *annectens* is uncommon, *townsendi* is rare, and *fuliginosa* is absent. Swarth inferred that those northern subspecies make an ocean crossing on their way from Alaska to California, much like the Aleutian Cackling Goose. Fraser et al. (2018) questioned this, but DeCicco et al. (2017) found large numbers of Fox Sparrows using

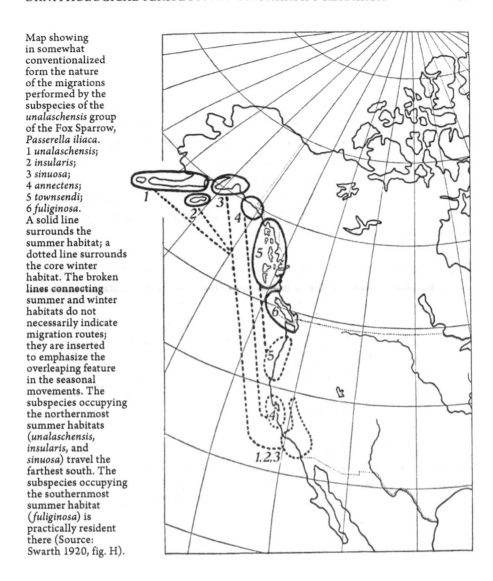

Map showing in somewhat conventionalized form the nature of the migrations performed by the subspecies of the *unalaschensis* group of the Fox Sparrow, *Passerella iliaca*. 1 *unalaschensis*; 2 *insularis*; 3 *sinuosa*; 4 *annectens*; 5 *townsendi*; 6 *fuliginosa*. A solid line surrounds the summer habitat; a dotted line surrounds the core winter habitat. The broken lines connecting summer and winter habitats do not necessarily indicate migration routes; they are inserted to emphasize the overleaping feature in the seasonal movements. The subspecies occupying the northernmost summer habitats (*unalaschensis*, *insularis*, and *sinuosa*) travel the farthest south. The subspecies occupying the southernmost summer habitat (*fuliginosa*) is practically resident there (Source: Swarth 1920, fig. H).

Middleton Island in the Gulf of Alaska as a stopover during fall migration, confirming Swarth's inference after all. A century after the publication of his landmark paper, Harry Swarth is still a vital participant in our never-ending conversation about the wonders of bird migration.

The Timberline Sparrow: An Incipient Species

BY CHRISTOPHER W. SWARTH

By the 1920s the discovery of a new breeding bird species north of Mexico was a significant ornithological event. Such an event took place on 8 July 1924. Swarth and Allan Brooks had been exploring the mountains a few miles south of Atlin, British Columbia, when they caught sight of two small, elusive sparrows that quickly hid in dense shrubbery. They managed to collect one bird that at first glance closely resembled a Brewer's Sparrow (*Spizella breweri*), a species completely unknown in the region. The entry in Swarth's notebook for 8 July 1924, registers their excitement:

> The surprise of the day—the greatest of the whole trip, thus far, for that matter—came when Brooks shot a Brewer's Sparrow. This was on the summit, well above timberline (1,463 m [4,800 ft] and higher), in the same surroundings as the Golden-crowned Sparrow. He saw two of the birds and shot one, in a thicket of balsam and birch. The one secured is an adult female, from the extensively denuded abdomen it is obviously a bird that had nested recently. This is a most extraordinary extension of range, from extreme southern British Columbia.

This record was hundreds of miles north of the known range of Brewer's Sparrow. Atlin is 1,450 km (900 mi) from the arid valleys of southern British Columbia and west-central Alberta, where the species was known to occur (American Ornithologists' Union 1910). To add to the puzzle, the Atlin sparrows inhabited steep mountain slopes covered with stunted willow and birch thickets just below tree line—habitat very different from the sagebrush-covered plains of the Great Basin, which are the Brewer's Sparrow's "center of abundance" (Swarth and Brooks 1925). The men spent six more days searching the high ridgetops and collecting a total of eighteen specimens. As they examined them in the hand, they were becoming convinced the birds were not Brewer's at all. Swarth's field notebook entry for 27 July 1924, reveals his uncertainty:

> I shot what I took to be a young Tree Sparrow, and picked up an adult Brewer's Sparrow, our second, if they really are Brewer's.

When Swarth returned to the Museum of Vertebrate Zoology (MVZ) in Berkeley he compared his and Brooks's Atlin specimens with the many

Brewer's Sparrow specimens already in the MVZ collection. After making careful measurements he found consistent differences in appearance between the two groups of birds. Compared with the large array of Brewer's Sparrow specimens collected from throughout the West, the Atlin sparrows were larger, with a longer tail; were darker and grayer overall, with heavier dorsal streaking; had a more prominent superciliary; and had a smaller, more slender, darker bill (Swarth and Brooks 1925; Doyle 1997). Swarth and Brooks were now convinced that their birds represented a new species, and they named it the Timberline Sparrow (*Spizella taverneri*). The common name indicates the habitat and elevation where it dwells, above the "upright" trees in the alpine-Arctic zone at elevations of 1,066 to 1,676 m (3,500 to 5,500 ft). The genus *Spizella* includes six species of small, slim sparrows with long, notched tails and small bills. The specific epithet *taverneri* honors Percy Algernon Taverner, ornithologist at the National Museum of Canada. Swarth and Brooks published their formal scientific description of the Timberline Sparrow in 1925 in the *Condor* (volume 27). The specimen designated as the holotype (a single type specimen on which the scientific description and name of a new species is based) was a male (MVZ 44856), collected on 8 August 1924, on Spruce Mountain, ten miles east of Atlin.

In 1926, Swarth and Brooks published their *Report on a Collection of Birds and Mammals from the Atlin Region, Northern British Columbia*, writing about the Timberline Sparrow:

> The discovery of this species was one of the most interesting of the season's results. . . . We first encountered the timber-line sparrow on July 8 (1924), near the summit of Monarch Mountain, about 4,500 feet altitude. The surroundings there are such as obtain generally above timber line in this region, the country being open, grass covered for the most part, the damper portions with small areas of false heather and the whole interspersed with clumps of scrubby balsam, mostly prostrate, but sometimes ten or fifteen feet high. It was a raw day, with showers at frequent intervals, the rain driving before a sharp wind, conditions such as to render a search for small birds difficult and unproductive. We were following a flock of horned larks when two sparrows appeared, perched upon a balsam thicket some distance away and jerking their tails nervously. Their appearance did not accord with anything we knew in the region, and Brooks started at once in pursuit. With some difficulty, for the birds were wary, he secured one of them.

Swarth and Brooks's conclusions were met with criticism by some. Chief among critics was Joseph Grinnell, director of the MVZ and Swarth's supervisor at the museum. In *Vertebrate Natural History of a Section of Northern California through the Lassen Peak Region* (1930), Grinnell and coauthors remarked offhandedly and based on examination of a single Brewer's Sparrow specimen from Lassen (the "Manzanita Lake bird"):

> We were astonished to find that our Manzanita Lake bird (an adult male, #48531) approximated or even duplicated *Spizella taverneri* in most of its peculiarities of coloration [see Swarth and Brooks 1925:67]. . . . But enough facts have come to light to suggest a "tendency" in birds from the extreme northwestern outposts in the general range of *breweri*, toward *taverneri*. Indeed, if the criterion of intergradation through individual variation be resorted to, the heretofore supposed gap between *breweri* and *taverneri* disappears, so that the two forms would best be treated as subspecies of one species.

Grinnell presented no "facts," but his opinion carried great weight with the American Ornithologists' Union (AOU) Check-list Committee, which was preparing to publish the fourth edition of the *AOU Check-list of North American Birds*. In October of that same year Grinnell had received a sparrow specimen that MVZ mammalogist Seth Benson had collected in New Mexico. Grinnell (1932) quickly published a note on this migrating sparrow, arguing again against species status for *taverneri*. By writing this he became "the first formally to merge *S. taverneri* with *S. breweri* and to provide a rationale (apparent intergradation of characteristics in birds of the northern United States) for doing this." (Banks and Gibson 2007). When the *AOU Check-list* was published in 1931, the Timberline Sparrow was not given full species status but was listed as a subspecies of the Brewer's Sparrow. Thus, the Timberline Sparrow's status as a full species lasted only from 1925 to 1931. (This was not entirely unusual. In the process of describing and naming a newly discovered bird, authors may initially assign it the rank of species. Later, new studies may result in a change in rank to subspecies). Swarth was not swayed by the AOU's decision and stood by his determination. He collected thirty-eight more sparrows from near Atlin in 1929, 1931, and 1934 and placed them in the collections at the California Academy of Sciences, and in his posthumously published *Birds of Atlin* (1936b), he continued to refer to the sparrow as *Spizella taverneri*. After Swarth passed away in 1935, the Timberline Sparrow controversy faded, but the story did not end.

Timberline Sparrow, Jasper
National Park, Alberta, July
2019. Photograph by Gavin
McKinnon.

Over the next sixty years the known range of the Timberline Sparrow expanded as birds were found breeding in new locations in Canada and nearby in Alaska. In 1945, one was heard singing at Teepee Lake in the St. Elias Mountains, Yukon Territory (Clarke 1945). Others were found in southwestern Alberta, 150 km (93 miles) north of the northern limit of where Brewer's were known to nest (McTaggert-Cowan 1946). In the 1990s the question of the status of the Timberline Sparrow arose again when new populations were discovered. In 1992 the range was found to extend across the Canadian border into the Gold Hill region of Alaska's Nutzotin Mountains and the Upper Cheslina drainage of the Mentasta Mountains (Doyle 1997). Terry Doyle, a biologist with the Tetlin National Wildlife Refuge in Alaska, investigated the sparrow's occurrence and biology over a five-year period and discovered eighteen birds, including many singing males and an adult feeding a recently fledged juvenile (Doyle 1997). He recorded their songs, noting that these were considerably different from the typical song of Brewer's Sparrow. He also compared his specimens with Brewer's, demonstrating that Timberline Sparrows were consistently larger. Adding to the story, in 1998 and 1999 Timberline Sparrows were found breeding much farther south at high elevations in Glacier National Park, Montana, not far from where Brewer's Sparrows nested (Griffin et al. 2003). This finding was potentially significant, because if the ranges of subspecies come into contact, questions about interbreeding can eventually be answered. Nonetheless, it was becoming well established that the two sparrow groups differed consistently in morphology, song, and breeding habitat, if not also in geographic range.

At this same time, the new field of biomolecular systematics was being applied to the Timberline Sparrow issue by Robert Zink at the University of Minnesota. Zink studies the genes and base pairs in mitochondrial DNA (mtDNA) to examine the relationships and evolutionary history of birds. He is a proponent of the phylogenetic species concept, which uses distinguishable differences among bird groups, either morphological or genetic, to reveal species-level relationships. Zink can be described as a splitter—he advocates for more species—as opposed to a lumper, who wants to combine species. A few years ago, Zink and colleagues (Barrowclough et al. 2016) stirred a controversy when they proposed that there may be eighteen thousand bird species worldwide rather than ten thousand, the widely accepted figure. Their estimate carries species splitting to a new level.

In 1999, Zink, John Klicka (also at the University of Minnesota), Terry Doyle, and two Canadian ornithologists, Jon Barlow and W. Bruce McGillivray, teamed up to investigate the taxonomic status and evolutionary history of the Timberline and Brewer's Sparrows. They compared dozens of skeletons, bird skins, and a large set of genetic data from both groups of birds. The Timberline Sparrow was found to be consistently larger than Brewer's, which supported the view of Swarth and Brooks (1926) and the more recent work by Doyle (1997). By examining mtDNA from feathers, skin, and heart and pectoral muscle and by using a molecular clock model, the team estimated that the Timberline and Brewer's Sparrows diverged in the late Pleistocene epoch and that the Timberline Sparrow is an incipient species that probably diverged before the last major glacial advance in the Pacific Northwest. The two sparrow groups do not come into contact during the breeding season, so the authors could not determine whether the Timberline and Brewer's are reproductively isolated—a key requirement of the Biological Species Concept, which defines species as "groups of freely interbreeding natural populations that are reproductively isolated from other such groups." Nonetheless, the researchers are convinced that the two sparrows are diagnosable evolutionary units that are evolving independently of one another. (Klicka commented later that in the cytochrome b gene [a diagnostic genetic marker], the Timberline and Brewer's Sparrows differ by only a single genetic mutation). Klicka, Zink, and colleagues concluded that "the data presented corroborate plumage, vocal, and ecological evidence suggesting that these taxa are distinct. As such, we suggest that *Spizella taverneri* be recognized as a species."

The team's 1999 paper prompted a response by evolutionary biologists Ernst Mayr of Harvard's Museum of Comparative Zoology and MVZ's Ned

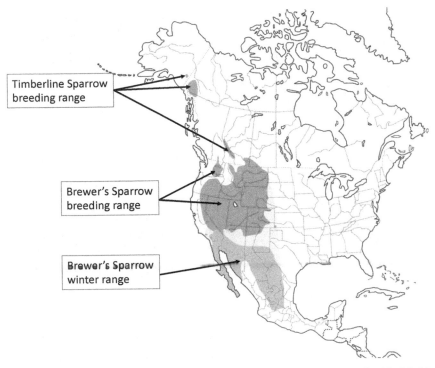

Breeding ranges of the Timberline Sparrow and Brewer's Sparrow. Source: *Birds of the World*, Cornell Lab of Ornithology, Ithaca, New York.

Johnson (Mayr and Johnson 2001). Mayr was a leader in the field of evolutionary biology in the twentieth century and the creator and leading proponent of the Biological Species Concept. He and Johnson were not convinced by the conclusions of Klicka and colleagues and argued in strong words for retaining the Timberline as a subspecies. Their main point was that it is impossible to know whether the two groups could interbreed because they are not sympatric (overlapping in distribution); their ranges are separated by more than 150 km (95 mi) (Starzomski 2015). When the extent of interbreeding is not known, populations of similar-looking, related birds that do not occur in the same area are considered by ornithologists to be the same species. In their response to Mayr and Johnson, Klicka et al. (2001) restated their case and concluded, "*Spizella taverneri* represents a very young species, displaying precisely the morphological and genetic characteristics we would expect to see in a newly evolved species. Indeed, we think it is one of the most likely examples of a Late Pleistocene speciation event in a North American bird." The AOU Check-list Committee, the final arbiter on nomenclature and classification, was not persuaded, and for the second time the Timberline Sparrow was denied full species status.

Brewer's Sparrow, Mono County, California. Photograph by Mark A. Chappell.

After the imbroglio between Klicka-Zink and Mayr-Johnson, Ned Johnson and Carla Cicero (2004) published a review paper examining the genetic makeup and Pleistocene epoch origin of thirty-nine pairs of proven sister bird species. Sister species are the descendant species formed when one species splits into two during evolution. Both sisters share an ancestral species not shared by any other species. *Spizella taverneri* and *S. breweri* are such sisters. In their analysis, Johnson and Cicero examined the Timberline Sparrow and Brewer's Sparrow data from Klicka et al. (1999). This time, the conclusions of Klicka and colleagues were not disputed; Johnson and Cicero agreed with them. In a turnaround from the paper by Mayr and Johnson (2001), Johnson and Cicero stated that the Timberline Sparrow is "borderline" and "probably best regarded as a species." Of all the pairs of sister species they examined, the Timberline Sparrow was the youngest, with the split occurring about thirty-five thousand years ago.

I could be accused of having a vested interest in this matter, but I favor treating the Timberline Sparrow as a full species. New molecular techniques will continue to provide deeper insight into the evolutionary relationship of these sparrows. Categorizing organisms will always remain an imperfect endeavor, and the concept of exactly what constitutes a species will continue to evolve, but Swarth and Brooks's discovery a century ago on the barren mountain slopes of British Columbia remains a singular event in North American ornithology.

SECTION 5

Glossary of Swarth's English and Scientific Names

Nomenclature of the 1930s is brought up to date (2021) here, including minor spelling changes, as in hyphenated names and possessive English names. English names for subspecies are included. The arrangement and order of species follow guidelines of the American Ornithological Society's North American Classification Committee.

White-cheeked Goose = Canada Goose (*Branta canadensis occidentalis*)
Hutchins' Goose = Cackling Goose (*Branta hutchinsii*)
Shoveler = Northern Shoveler (*Spatula clypeata*)
Baldpate = American Wigeon (*Mareca americana*)
Pintail = Northern Pintail (*Anas acuta*)
Oldsquaw = Long-tailed Duck (*Clangula hyemalis*)
American Golden-eye = Common Goldeneye (*Bucephala americana*)
Barrow's Golden-eye = Barrow's Goldeneye (*Bucephala islandica*)
American Merganser = Common Merganser (*Mergus merganser americanus*)

Willow Grouse = Ruffed Grouse (*Bonasa umbellus*)
Gray Ruffed Grouse = Ruffed Grouse (*Bonasa umbellus yukonensis*)
Oregon Ruffed Grouse = Ruffed Grouse (*Bonasa umbellus sabini*)
Valdez Spruce Grouse = Spruce Grouse (*Canachites canadensis canadensis*) (incl. subspecies *atratus*)
Franklin's Grouse = Spruce Grouse (*Canachites canadensis franklinii*) [British Columbia] and (*F. c. isleibi*) [Alaska]
Willow Ptarmigan (*Lagopus lagopus albus*) = *Lagopus lagopus alba*
Rock Ptarmigan (*Lagopus mutus*) = *Lagopus muta*
Lagopus mutus mutus = *Lagopus muta muta*
Lagopus rupestris = *Lagopus muta*
Lagopus rupestris rupestris = *Lagopus muta rupestris*
White-tailed Ptarmigan (*Lagopus leucurus*) = *Lagopus leucura*
Blue Grouse = Sooty and Dusky Grouse together as a single species
Sitka Grouse = Sooty Grouse (*Dendragapus fuliginosus sitkensis*)
Fleming's Grouse = Dusky Grouse (*Dendragapus obscurus flemingi*)

Holboell's Grebe = Red-necked Grebe (*Podiceps grisegena holboellii*)

Eastern Nighthawk = Common Nighthawk (*Chordeiles minor minor*)

Vaux Swift = Vaux's Swift (*Chaetura vauxi*)

Coot = American Coot (*Fulica americana*)
Little Brown Crane = Sandhill Crane (*Antigone canadensis canadensis*)

American Golden Plover = American Golden-Plover (*Pluvialis dominica*)

Bartramia Sandpiper = Upland Sandpiper (*Bartramia longicauda*)
Hudsonian Curlew = Whimbrel (*Numenius phaeopus hudsonicus*)
Surf-bird = Surfbird (*Calidris virgata*)
Red-backed Sandpiper = Dunlin (*Calidris alpina*)
Aleutian Sandpiper = Rock Sandpiper (*Calidris ptilocnemis*)
Baird Sandpiper = Baird's Sandpiper (*Calidris bairdii*)
Northern Phalarope = Red-necked Phalarope (*Phalaropus lobatus*)

California Murre = Common Murre (*Uria aalge inornata*)

Short-billed Gull = Mew Gull (*Larus canus brachyrhynchus*)

Great Northern Diver = Common Loon (*Gavia immer*)
immer = Common Loon (*Gavia immer*)
Fork-tailed Petrel = Fork-tailed Storm-Petrel (*Hydrobates furcatus*)
Beal's Petrel = Leach's Storm-Petrel (*Hydrobates leucorhous leucorhous*) (incl. subspecies *bealei*)

White-crested Cormorant = Double-crested Cormorant (*Nannopterum auritum cincinatum*)

Northern Bald Eagle = Bald Eagle (*Haliaeetus leucocephalus*)
Kamchatka Sea Eagle = Steller's Sea-Eagle (*Haliaeetus pelagicus*)
Marsh Hawk = Northern Harrier (*Circus hudsonius*)
Goshawk = Northern Goshawk (*Accipiter gentilis*)
Western Red-tailed Hawk = Red-tailed Hawk (*Buteo jamaicensis calurus*) (and/or subspecies *alascensis*)
Harlan's Hawk = Red-tailed Hawk (*Buteo jamaicensis harlani*)

Horned Owl = Great Horned Owl (*Bubo virginianus*)
Arctic Horned Owl = Great Horned Owl (*Bubo virginianus subarcticus*)
Dusky Horned Owl = Great Horned Owl (*Bubo virginianus saturatus*)
Great Gray Owl (*Scotiaptex nebulosa nebulosa*) = Great Gray Owl (*Strix nebulosa nebulosa*)
Hawk Owl = Northern Hawk Owl (*Surnia ulula*)
Pygmy Owl = Northern Pygmy-Owl (*Glaucidium gnoma*)
Acadian Owl = Northern Saw-whet Owl (*Aegolius acadicus acadicus*)
Fleming's Owl (*Cryptoglaux flemingi*) = Northern Saw-whet Owl (*Aegolius acadicus brooksi*)
Richardson's Owl = Boreal Owl (*Aegolius funereus richardsoni*)

Western Belted Kingfisher (*Ceryle alcyon caurina*) = Belted Kingfisher (*Megaceryle alcyon*)

Alaskan Three-toed Woodpecker = American Three-toed Woodpecker (*Picoides dorsalis fasciatus*) (incl. subspecies *alascensis* and *fumipectus*)
Nelson's Downy Woodpecker = Downy Woodpecker (*Dryobates pubescens nelsoni*)
Rocky Mountain Downy Woodpecker = Downy Woodpecker (*Dryobates pubescens leucurus*)
Gairdner's Woodpecker = Hairy Woodpecker (*Dryobates villosus gairdneri*)
Harris Woodpecker = Hairy Woodpecker (*Dryobates villosus harrisi*)
Rocky Mountain Hairy Woodpecker = Hairy Woodpecker (*Dryobates villosus monticola*)
Sitka Hairy Woodpecker = Hairy Woodpecker (*Dryobates villosus sitkensis*)
Yellow-shafted Flicker = Northern Flicker (*Colaptes auratus luteus*)
Red-shafted Flicker = Northern Flicker (*Colaptes auratus cafer*)

Sparrow Hawk = American Kestrel (*Falco sparverius*)
Pigeon Hawk = Merlin (*Falco columbarius*)
Black Merlin = Merlin (*Falco columbarius suckleyi*)
Duck Hawk = Peregrine Falcon (*Falco peregrinus anatum*)
Peale's Falcon = Peregrine Falcon (*Falco peregrinus pealei*)

Western Wood Pewee = Western Wood-Pewee (*Contopus sordidulus*)
Western Flycatcher = Pacific-slope Flycatcher (*Empidonax difficilis difficilis*)
Wright's Flycatcher = Dusky Flycatcher (*Empidonax oberholseri*)

Western Warbling Vireo = Warbling Vireo (*Vireo gilvus swainsoni*)

Steller Jay = Steller's Jay (*Cyanocitta stelleri*)
Black-headed Jay = Steller's Jay (*Cyanocitta stelleri annectens*)
Magpie = Black-billed Magpie (*Pica hudsonia*)
Western Crow = American Crow (*Corvus brachyrhynchos hesperis*)
Beach Crow = American Crow (*Corvus brachyrhynchos caurinus*)
Northwest-Crow/Northwest Crow = American Crow (*Corvus brachyrhynchos caurinus*)
Raven = Common Raven (*Corvus corax*)

Pallid Horned Lark (*Otocoris alpestris arcticola*) = Horned Lark (*Eremophila alpestris arcticola*)

Rough-winged Swallow = Northern Rough-winged Swallow (*Stelgidopteryx serripennis*)

Long-tailed Chickadee = Black-capped Chickadee (*Poecile atricapillus septentrionalis*)
Penthestes atricapillus = *Poecile atricapillus*
Hudsonian Chickadee = Boreal Chickadee (*Poecile hudsonicus*)

Canada Nuthatch = Red-breasted Nuthatch (*Sitta canadensis*)

Rocky Mountain Creeper = Brown Creeper (*Certhia familiaris montana*)
Tawny Creeper = Brown Creeper (*Certhia familiaris occidentalis*)

Western House Wren = House Wren (*Troglodytes aedon parkmani*)
Western Winter Wren = Pacific Wren (*Troglodytes pacificus pacificus*)

Dipper = American Dipper (*Cinclus mexicanus*)
Water Ouzel = American Dipper (*Cinclus mexicanus*)

Sitka Ruby-crowned Kinglet = Ruby-crowned Kinglet (*Corthylio calendula grinnelli*)
Eastern Ruby-crowned Kinglet = Ruby-crowned Kinglet (*Corthylio calendula calendula*)
Western Golden-crowned Kinglet = Golden-crowned Kinglet (*Regulus satrapa olivaceus*)

Rocky Mountain Bluebird = Mountain Bluebird (*Sialia currucoides*)
Townsend Solitaire = Townsend's Solitaire (*Myadestes townsendi*)
Olive-backed Thrush = Swainson's Thrush (*Catharus ustulatus swainsoni*)
Hylocichla ustulata = *Catharus ustulatus*
Russet-backed Thrush = Swainson's Thrush (*Catharus ustulatus ustulatus*)
Eastern Hermit Thrush (*Hylocichla guttata pallasii*) = Hermit Thrush (*Catharus guttatus faxoni*)
Alaskan Hermit Thrush = Hermit Thrush (*Catharus guttatus guttatus*)
Dwarf Hermit Thrush = Hermit Thrush (*Catharus guttatus nanus*)

Robin = American Robin (*Turdus migratorius*)
Eastern Robin = American Robin (*Turdus migratorius migratorius*)
Coast Varied Thrush = Varied Thrush (*Ixoreus naevius naevius*)
Northern Varied Thrush = Varied Thrush (*Ixoreus naevius meruloides*)

Catbird = Gray Catbird (*Dumetella carolinensis*)
Cedar Bird = Cedar Waxwing (*Bombycilla cedrorum*)
Pipit = American Pipit (*Anthus rubescens*)
Rosy Finch = Gray-crowned Rosy-Finch (*Leucosticte tephrocotis*)
Leucosticte = Gray-crowned Rosy-Finch (*Leucosticte tephrocotis*)
Hepburn's Rosy Finch = Gray-crowned Rosy-Finch (*Leucosticte tephrocotis littoralis*)
Eastern Purple Finch = Purple Finch (*Haemorhous purpureus purpureus*)
Bendire's Crossbill = Red Crossbill (*Loxia curvirostra bendirei*)
Redpoll = Common Redpoll (*Acanthis flammea*)

Alberta Fox Sparrow = Fox Sparrow (*Passerella iliaca altavagans*)
Sooty Fox Sparrow = Fox Sparrow (*Passerella iliaca fuliginosa*)
Townsend's Fox Sparrow = Fox Sparrow (*Passerella iliaca townsendi*)
Oregon Junco = Dark-eyed ("Oregon") Junco (*Junco hyemalis oreganus*)
Cassiar Junco = Dark-eyed ("Slate-colored") Junco (*Junco hyemalis cismontanus*)
Gambel's Sparrow = White-crowned Sparrow (*Zonotrichia leucophrys gambelii*)
Aleutian Savannah Sparrow = Savannah Sparrow (*Passerculus sandwichensis sandwichensis*)
Western Savannah Sparrow = Savannah Sparrow (*Passerculus sandwichensis alaudinus*)
Forbush's Lincoln Sparrow = Lincoln Sparrow (*Melospiza lincolnii gracilis*)
Rusty Song Sparrow = Song Sparrow (*Melospiza melodia morphna*)
Sooty Song Sparrow = Song Sparrow (*Melospiza melodia rufina*)

Grinnell's Water-thrush = Northern Waterthrush (*Parkesia noveboracensis*)

Lutescent Warbler = Orange-crowned Warbler (*Leiothlypis celata lutescens*)

Western Orange-crowned Warbler = Orange-crowned Warbler (*Leiothlypis celata orestera*)

Eastern Orange-crowned Warbler = Orange-crowned Warbler (*Leiothlypis celata celata*)

Tolmie Warbler = MacGillivray's Warbler (*Geothlypis tolmiei*)

Western Yellowthroat = Common Yellowthroat (*Geothlypis trichas occidentalis*)

Redstart = American Redstart (*Setophaga ruticilla*)

Eastern Yellow Warbler = Yellow Warbler (*Setophaga petechia aestiva*)

Alaska Yellow Warbler = Yellow Warbler (*Setophaga petechia rubiginosa*)

Myrtle (Hoover's) Warbler = Yellow-rumped Warbler (*Setophaga coronata hooveri*)

Audubon Warbler = Yellow-rumped Warbler (*Setophaga coronata auduboni*)

nigriferous = *Setophaga coronata auduboni*

coronata = *Setophaga coronata auduboni*

Louisiana Tanager = Western Tanager (*Piranga ludoviciana*)

SECTION 6

Publications by Harry S. Swarth on the Birds and Mammals of Northwestern British Columbia and Southeastern Alaska

(key publications are in **bold**)

Review of Osgood's "Biological investigations in Alaska and Yukon Territory." 1909. *Condor* 11:209–210.

Birds and mammals of the 1909 Alexander Alaska Expedition. 1911. *University of California Publications in Zoology* **7:9–172.**

Description of a new hairy woodpecker from southeastern Alaska. 1911. *University of California Publications in Zoology* 7:313–318.

Review of Oberholser's "A revision of the forms of the Hairy Woodpecker (*Dryobates villosus* [Linnaeus])." 1911. *Condor* 13:169–170.

Notes from Alaska. 1911. *Condor* 13:211.

A visit to Nootka Sound. 1912. *Condor* 14:15–21.

Report on a collection of birds and mammals from Vancouver Island. 1912. *University of California Publications in Zoology* **10:1–124.**

The winter range of the Yakutat song sparrow. 1912. *Condor* 14:73.

Review of Riley's "Birds collected or observed on the expedition of the Alpine Club of Canada to Jasper Park, Yellowhead Pass, and Mount Robson Region." 1913. *Condor* 15:130–131.

Review of Oberholser's "A revision of the forms of the great blue heron (*Ardea herodias* Linnaeus)." 1913. *Condor* 15:50–51.

Review of E. M. Anderson's "Report on birds collected and observed during April, May, and June, 1913, in the Okanagan Valley, from Okanagan Landing south to Osoyoos Lake," and "Birds collected and observed in the Atlin District, 1914," and F. Kermode's and E. M. Anderson's "Report of birds collected and observed during September, 1913, on Atlin Lake, from Atlin to south end of the lake." 1915. *Condor* 17:133–134.

Review of J. A. Munro's "Report of field-work in Okanagan and Shuswap districts, 1916." 1918. *Condor* 20:48.

The subspecies of the Oregon jay. 1918. *Condor* 20:83–84.

Review of J. Dwight's "The geographic distribution of color and of other variable characters in the genus *Junco*—a new aspect of specific and subspecific values." 1918. *Condor* 20:142–143.

Three new subspecies of *Passerella iliaca*. 1918. *Proceedings of the Biological Society of Washington* 31:161–164.

Revision of the avian genus *Passerella*, with special reference to the distribution and migration of the races in California. 1920. *University of California Publications in Zoology* 21:75–224.

The subspecies of *Branta canadensis* (Linnaeus). 1920. *Auk* 37:268–272.

The Sitkan race of the dusky grouse. 1921. *Condor* 23:59–60.

The red squirrel of the Sitkan district, Alaska. 1921. *Journal of Mammalogy* 2:92–94.

Birds and mammals of the Stikine River region of northern British Columbia and southeastern Alaska. 1922. *University of California Publications in Zoology* 24:125–314.

The Bohemian waxwing: A cosmopolite. 1922. *University of California Chronicle* 24:450–455.

The systematic status of some northwestern song sparrows. 1923. *Condor* 25:214–223.

Birds and mammals of the Skeena River region of northern British Columbia. 1924. *University of California Publications in Zoology* 24:315–394.

The timberline sparrow, a new species from northwestern Canada. 1925. *Condor* 27:67–69 (with Allan Brooks).

A visit to the Stikine glaciers. 1925. *Sierra Club Bulletin* 12:121–125.

A Distributional List of the Birds of British Columbia. 1925. *Pacific Coast Avifauna* 17. 158 pp. (with Allan Brooks).

Report on a collection of birds and mammals from the Atlin region, northern British Columbia. 1926. *University of California Publications in Zoology* 30:51–162.

Review of P. A. Taverner's "Birds of western Canada." 1927. *Condor* 29:84–85.

Eversmann shrike not a North American bird. 1927. *Condor* 29:205.

Birds of the Atlin region, British Columbia: A reply to criticism. 1927. *Condor* 29:169–170.

The rufous-necked sandpiper on St. Paul, Pribilof Islands. 1927. *Condor* 29:200–201.

The rufous-necked sandpiper in Alaska. 1927. *Condor* 29:274.

Review of Taverner's "A study of *Buteo borealis*, the red-tailed hawk, and its varieties in Canada." 1928. *Condor* 30:197–199.

Occurrence of some Asiatic birds in Alaska. 1928. *Proceedings of the California Academy of Sciences* 17:247–251.

Notes on the avifauna of the Atlin region, British Columbia. 1930. *Condor* 32:216–217.

Nesting of the timberline sparrow. 1930. *Condor* 32:255–257.

Geographic variation in the Richardson grouse. 1931. *Proceedings of the California Academy of Sciences* 20:1–7.

The lemming of Nunivak Island, Alaska. 1931. *Proceedings of the Biological Society of Washington* 44:101–104.

The long-tailed meadow-mouse of southeastern Alaska. 1933. *Proceedings of the Biological Society of Washington* 46:207–212.

Off-shore migrants over the Pacific. 1933. *Condor* 35:39–41.

Problems in the classification of northwestern horned owls. 1934. *Condor* 36:38–40.

Birds of Nunivak Island, Alaska. 1934. *Pacific Coast Avifauna* 22:1–64.

A barn swallow's nest on a moving train. 1935. *Condor* 37:84–85.

Review of E. R. Hall's "Mammals collected by T. T. and E. B. McCabe in the Bowron Lake region of British Columbia." 1935. *Canadian Field-Naturalist* 49:77–78.

Systematic status of some northwestern birds. 1935. *Condor* 37:199–204.

Savannah sparrow migration routes in the northwest. 1936. *Condor* 38:30–32.

A list of the birds of the Atlin region, British Columbia. 1936. *Proceedings of the California Academy of Sciences* 23:35–58.

Origins of the fauna of the Sitkan district, Alaska. 1936. *Proceedings of the California Academy of Sciences* 23:59–71.

An early estimate of California's fauna. 1936. *Condor* 38:38–39.

Mammals of the Atlin region, northwestern British Columbia. 1936. *Journal of Mammalogy* 17:398–405.

LITERATURE CITED

American Ornithologists' Union. *AOU Check-list of North American Birds*. 1910. 3rd ed. New York: American Ornithologists' Union.

————. *AOU Check-list of North American Birds*. 1931. 4th ed. Lancaster, PA: American Ornithologists' Union.

————. *AOU Check-list of North American Birds*. 1998. 7th ed. Lawrence, KS: Allen Press.

Bailey, A. M. 1927. Notes on the birds of southeastern Alaska. *Auk* 44:184–205.

Banks, R. C., and D. D. Gibson. 2007. The correct type locality of *Spizella breweri*. *Auk* 124:1083–1085.

Barrowclough, G. F., J. Cracraft, J. Klicka, and R. M. Zink. 2016. How many kinds of birds are there and why does it matter? *PLoS ONE* 11(11): e0166307. https://doi.org/10.1371/journal.pone.0166307.

Barry, P. D., and D. A. Tallmon. 2010. Genetic differentiation of a subspecies of Spruce Grouse (*Falcipennis canadensis*) in an endemism hotspot. *Auk* 127:617–625.

Bell, C. P. 1997. Leap-frog migration in the Fox Sparrow: Minimizing the cost of spring migration. *Condor* 99:470–477.

Boland, J. M. 1990. Leapfrog migration in North American shorebirds: Intra- and interspecific examples. *Condor* 92:284–290.

Brooks, A., and H. S. Swarth. 1925. *A Distributional List of the Birds of British Columbia*. Pacific Coast Avifauna 17.

Campbell, W. R., N. K. Dawe, I. McTaggart-Cowan, J. M. Cooper, G. W. Kaiser, M. C. E. McNall, G. E. Smith, and A. C. Stewart. 1990–2001. *The Birds of British Columbia*. Vols. 1–4. Vancouver: University of British Columbia Press.

Clarke, C. H. D. 1945. Some bird records from Yukon Territory. *Canadian Field-Naturalist* 59:65.

Cook, J. A., N. G. Dawson, and S. O. MacDonald. 2006. Conservation of highly fragmented systems: The north temperate Alexander Archipelago. *Biological Conservation* 133:1–15.

Davidson, P. J. A., R. J. Cannings, A. R. Couturier, D. Lepage, and C. M. Di Corrado, eds. 2015. *Atlas of the Breeding Birds of British Columbia, 2008–2012.* Delta, BC: Bird Studies Canada. Accessed August 2021. http://www.birdatlas.bc.ca.

DeCicco, L. H., D. D. Gibson, T. G. Tobish Jr., S. C. Heinl, N. R. Hajdukovich, J. A. Johnson, and C. W. Wright. 2017. Birds of Middleton Island, a unique landfall for migrants in the Gulf of Alaska. *Western Birds* 48:214–293.

Dickerman, R. W., and J. Gustafson. 1996. The Prince of Wales Spruce Grouse: A new subspecies from southeastern Alaska. *Western Birds* 27:41–47.

Doyle, T. J. 1997. The Timberline Sparrow, *Spizella (breweri) taverneri*, in Alaska, with notes on breeding habitat and vocalizations. *Western Birds* 28:1–12.

Drent, R. H., and T. Piersma. 1989. An exploration of the energetics of leap-frog migration in Arctic breeding waders. In *Bird Migration*, edited by E. Gwinner, 399–412. Berlin: Springer-Verlag.

Elphick, J. 2007. *The Atlas of Bird Migration: Tracing the Great Journeys of the World's Birds.* Richmond Hill, Ontario: Firefly Books.

Fontaine, J. J., R. J. Stutzman, and L. Z. Gannes. 2015. Leaps, chains, and climate change for western migratory songbirds. In *Phenological Synchrony and Bird Migration: Changing Climate and Seasonal Resources in North America*, edited by E. M. Wood and J. L. Kellermann, 3–15. Studies in Avian Biology 47.

Fraser, K. C., A. Roberto-Charron, B. Cousens, M. Simmons, A. Nightingale, A. C. Shave, R. L. Cormier, and D. L. Humple. 2018. Classic pattern of leapfrog migration in Sooty Fox Sparrow (*Passerella iliaca unalaschcensis*) is not supported by direct migration tracking of individual birds. *Auk* 135:572–582. Gabrielson, I. N., and F. C. Lincoln. 1959. *Birds of Alaska.* Harrisburg, PA: Stackpole, Wildlife Management Institute.

Griffin, S. C., B. Walker, and M. M. Hart. 2003. Using GIS to guide field surveys for timberline sparrows in northwestern Montana. *Northwest Science* 77:54–63.

Grinnell, J. 1898. Summer birds of Sitka, Alaska. *Auk* 15:122–131.

———. 1909. Birds and mammals of the 1907 Alexander Expedition to southeastern Alaska. *University of California Publications in Zoology* 5:171–264.

———. 1932. A United States record of the Timberline Sparrow. *Condor* 34:231–232.

Grinnell, J., J. Dixon, and J. Linsdale. 1930. Vertebrate natural history of a section of northern California through the Lassen Peak region." *University of California Publications in Zoology* 35:1–594.

Heinl, S. C., and A. W. Piston. 2009. Birds of the Ketchikan area, southeastern Alaska. *Western Birds* 40:54–144.

Herman, S. G. 1986. *The Naturalist's Field Journal: A Manual of Instruction Based on a System Established by Joseph Grinnell.* Vermillion, SD: Buteo Books.

Howe, J. R., Jr. 1996. *The Bear Man of Admiralty Island: A Biography of Allen E. Hasselborg.* Fairbanks: University of Alaska Press.

Johnson, J. A., B. A. Andres, and J. A. Bisonette. 2008. *Birds of the Major Mainland Rivers of Southeast Alaska.* General Technical Report PNW-GTR-739. Portland, OR: US Forest Service Pacific Northwest Field Station.

Johnson, N. K., and C. Cicero. 2004. New mitochondrial DNA data affirm the importance of Pleistocene speciation in North American birds. *Evolution* 58:1122–1130.

Kelly, J. F., V. Atudoreim, Z. D. Sharp, and D. M. Finch. 2002. Insights into Wilson's Warbler migration from analyses of hydrogen stable-isotope ratios. *Oecologia* 130:216–221.

Kessel, B., and D. D. Gibson. 1978. *Status and Distribution of Alaska Birds*. Studies in Avian Biology 1.

———. 1994. A century of avifaunal change in Alaska. In *A Century of Avifaunal Change in Western North America*, edited by J. R. Jehl and N. K. Johnson, 4–13. Studies in Avian Biology 15.

Klicka, J., R. M. Zink, J. C. Barlow, W. B. McGillivray, and T. J. Doyle. 1999. Evidence supporting the recent origin and species status of the Timberline Sparrow. *Condor* 101:577–588.

———. 2001. The taxonomic rank of *Spizella taverneri*: A response to Mayr and Johnson. *Condor* 103:420–422.

Linsdale, J. M. 1936. Harry Schelwald Swarth. *Condor* 38:155–168.

Mailliard, J. 1937. In memoriam: Harry Schelwald Swarth, 1878–1935. *Auk* 54:127–134.

Mayr, E., and N. K. Johnson. 2001. Is *Spizella taverneri* a species or a subspecies?" *Condor* 103:418–419.

MacDonald, S., and J. A. Cook. 1996. The land mammal fauna of southeast Alaska. *Canadian Field-Naturalist* 110:571–598.

McTaggart-Cowan, I. 1946. Notes on the distribution of *Spizella breweri taverneri*. *Condor* 48:93–94.

Newton, I. 2007. *The Migration Ecology of Birds*. London: Academic Press.

Remsen, J. V., Jr. 1977. On taking field notes. *American Birds* 31:946–953.

Rosenberg, K. V., A. M. Dokter, P. J. Blancher, J. R. Sauer, A. C. Smith, P. A. Smith, J. C. Stanton, et al. 2019. Decline of the North American avifauna. *Science* 366:120–124.

Smith, M. A. 2016. *Ecological Atlas of Southeast Alaska*. Anchorage: Alaska Audubon.

Starzomski, B. 2015. Brewer's Sparrow. In *Atlas of the Breeding Birds of British Columbia, 2008–2012*, edited by P. J. A. Davidson, R. J. Cannings, A. R. Couturier, D. Lepage, and C. M. Di Corrado. Delta, BC: Bird Studies Canada. Accessed August 4, 2021. http://www.birdatlas.bc.ca/accounts/speciesaccount.jsp?sp=BRSP&lang=en.

Swarth, C. W. 2018. *An Expedition to Ramsey Canyon: The 1896 Field Journal of Ornithologist Harry S. Swarth*. Mariposa, CA: Yaqui Gulch Press.

Swarth, H. S. 1911. Birds and mammals of the 1909 Alexander Alaska Expedition. *University of California Publications in Zoology* 7:9–172.

———. 1914. *A Distributional List of the Birds of Arizona*. Cooper Ornithological Club, Pacific Coast Avifauna, No. 10.

———. 1920. Revision of the avian genus *Passerella*, with special reference to the distribution and migration of the races in California. *University of California Publications in Zoology* 21:75–224.

———. 1922. Birds and mammals of the Stikine River region of northern British Columbia and southeastern Alaska. *University of California Publications in Zoology* 24:125–314.

————. 1926. Report on a collection of birds and mammals from the Atlin region, northern British Columbia. *University of California Publications in Zoology* 30:51–162.

————. 1930. Nesting of the Timberline Sparrow. *Condor* 32:255–257.

————. 1931. *The Avifauna of the Galapagos Islands*. Occasional Papers of the California Academy of Sciences, No. 18.

————. 1934. The bird fauna of the Galapagos Islands in relation to species formation. *Biological Reviews* 9:213–234.

————. 1936a. Origins of the fauna of the Sitkan district, Alaska. *Proceedings of the California Academy of Sciences* 23:59–78.

————. 1936b. A list of the birds of the Atlin region, British Columbia. *Proceedings of the California Academy of Sciences* 24:35–58.

Swarth, H. S., and A. Brooks. 1925. The Timberline Sparrow, a new species from northwestern Canada. *Condor* 27:67–69.

————. 1926. Report on a collection of birds and mammals from the Atlin region, northern British Columbia. *University of California Publications in Zoology* 30:51–162.

Weckstein, J. D., D. E. Kroodsma, and R. C. Faucett. 2020. Fox Sparrow (*Passerella iliaca*), version 1.0. In *Birds of the World*, edited by A. F. Poole and F. B. Gill. Ithaca, NY: Cornell Lab of Ornithology. https://doi.org/10.2173/bow.foxspa.01.

Weckworth, B., S. Talbot, G. K. Sage, D. K. Person, and J. A. Cook. 2005. A signal for independent coastal and continental histories among North American wolves. *Molecular Biology* 14:917–931.

Welty, J. C., and L. Baptista. 1988. *The Life of Birds*. Philadelphia: W. B. Saunders.

Willett, G. 1915. Summer birds of Forrester Island, Alaska. *Auk* 32:295–305.

————. 1923. Bird records from Craig, Alaska. *Condor* 25:105–106.

————. 1927. Notes on the occurrences and distribution of some southeastern Alaskan birds. *Condor* 29:58–60.

————. 1928. Notes on some birds of southeastern Alaska. *Auk* 45:445–449.

Winker, K. 2010. Subspecies represent geographically partitioned variation, a gold mine of evolutionary biology, and a challenge for conservation. *Ornithological Monographs* 2010: 6–23.